This book is to be returned on or before
the last date stamped below.

NOV

MAR 1

2

Dynamics of
of
Petroleum Reservoirs
Under
Gas Injection

Dynamics
of
Petroleum Reservoirs
Under
Gas Injection

Rafael Sandrea

National Research Council
of Venezuela

Ralph F. Nielsen

Professor Emeritus
The Pennsylvania State University

Gulf Publishing Company
Book Division
Houston, Texas

Dedication

To Marietta and Elin

Library of Congress
Catalog Card Number
74-4829
ISBN 0-87201-219-0

Foreword

The main theme of this book is concerned with the technoeconomic aspects of gas injection, the stalwart method of pressure maintenance and secondary recovery in most oil producing countries. The book is an outgrowth of a short course on the same topic given by the authors to engineers from industry, in Venezuela, Trinidad and Mexico during the last five years. The original notes were broadened and used as part of the material taught in two graduate level courses at the Universidad Nacional Autonoma de Mexico.

In general the contents of this book have been heavily based on the publications of numerous investigators. The authors have tried to cull the literature so as to present a unified view of the subject matter, with a strong emphasis on engineering applications. For this reason, several problem examples are demonstrated for the reader and an extensive set of problems, based on well-known field case histories, are given in Appendix A.

Chapter 1 discusses the evolution of the gas displacement process and briefly describes a few classical field examples which show good engineering practice. Chapter 2 is concerned with a review of the basic tools of the engineer, namely, material balance techniques. In Chapters 3 and 4, the gas injection problem is handled from the dynamics of the displacement process, under immiscible and miscible conditions.

We are aware that many views expressed on several topics covered in this book may have been proposed by several authors to whom we may not have given due credit. We wish to apologize for this unintentional oversight. We would also like to thank Drs. S.M. Farouq Ali and Robert Banks for their many criticisms of the original manuscript, which definitely enhanced its readability.

R. Sandrea
R.F. Nielsen

Contents

3. Immiscible Gas-Oil Displacement   58.

4. Gas-Oil Displacement With Mass Transfer

Appendix A: Special Problem Sets

1. History of Gas Injection

1.0 Early Projects [16]

The injection of gas into an oil reservoir for the purpose of increasing oil recovery goes back to about 1890 in Venango County, Pennsylvania. In the earliest known case, an operator, James Dinsmoor, purposely allowed communication between a partially depleted oil sand and an underlying gas sand which had not been depleted. A few years later he installed "gas pumps," which pulled a partial vacuum on some wells while discharging the gas so obtained into other wells. About 1910, compressors were becoming common and air was often injected, either alone or mixed with the produced gas. Compression served the dual purpose of condensing casinghead gasoline and providing the pressure for reinjecting the gas into the reservoir. Gas and/or air injection was practiced in Pennsylvania and West Virginia before 1910 and, by 1911, projects had been started in Ohio. Within the next fifteen years, "repressuring" projects had been started in other states such as Kentucky, Illinois, Oklahoma, Kansas and Texas.

It should be noted that the early (prior to about 1930) repressuring projects were in highly depleted fields and therefore must be regarded as "secondary recovery" operations. Secondary recovery is generally defined as recovery by a method which augments the reservoir energy remaining after the reservoir has approached its economic production limits by the energy initially present. Also, in the early days the injection was at low pressures, usually under 100 psi, and into shallow formations. However, U.S. Patent No. 308,522, issued in 1884, mentions a pressure of 1000 psi. The use of a vacuum pump on the producing wells was quite common, partly because of the "richer" casinghead gasoline so produced, and partly because of at least a temporary increase in oil recovery.

Later it became apparent that the ultimate recovery was not appreciably increased, and vacuum pumps were largely abandoned where repressuring was in operation.

As mentioned previously, air and gas injection, or repressuring, was at first limited to secondary recovery operations. In the early days, location of injection wells relative to producing wells did not follow definite patterns, for some producing wells might be converted to input wells. The mathematics of fluid movement in the reservoir had not been developed to any extent before the 1920's. While Darcy's law dates from 1856[3], it had been used chiefly by soil and hydrology engineers. Oil producers were, however, aware of fluid movements in a general sense. The need for "offset wells" to keep oil from moving across lease lines was well recognized, as was the fact that injected gas would flow more or less radially toward surrounding producing wells.

The difficulties involved with individual lease operations became increasingly apparent, especially in cases where repressuring had been started. The first attempt to operate a large portion of a field as a unit was in Nowata County, Oklahoma. Since, however, records of oil production according to leases had to be kept, this was not strictly a unitization project. The first case of gas injection with unitization in the usual sense was in Kettleman Hills Reservoir in California. A number of criteria have been used to determine the returns to individual lease holders, such as acreage, productive sand volume, total well potentials, etc. With the rapidly increasing understanding of reservoir behavior during the late 1920's and early 1930's, it was apparent that different fields required different techniques for the most efficient operation, both from economic and conservation points of view. This will be discussed in detail later. To accomplish this, operating policies must be decided upon for each field as a whole, and this requires unitization.

Although Darcy's law and the concepts of porosity, permeability, and fluid properties were generally known previously, serious attempts to apply scientific methods to oil production date from the late 1920's[7]. During the next ten years the application of scientific methods grew very rapidly, particularly since the general methods had been known for a long time in other industries. While concepts of porosity and permeability had been applied in ceramic, soil, and hydrology engineering for decades, extensive coring of oil wells, with laboratory determinations of porosity, permeability, and fluid saturations, first began about 1930. Material balance methods, long known to chemical engineers, really began with the publication of a reservoir material balance equation by Schilthuis[15].

The analogy between Darcy's law and Fourier's and Ohm's laws of heat and electrical conduction, and similarity of continuity equations, enabled Muskat and others[13,14] to rapidly develop equations for fluid movements and pressure distributions in porous media. Some of the emphasis on "capillary retention" and "Jamin effect" disappeared when the concepts of relative permeability and "residual saturations" appeared with the work of Wyckoff and Botset[20] in 1936 and that of Leverett[10] in 1939. Displacement of one phase by another in porous media was first put on a scientific basis with the application of continuity and flow equations to separate phases by Buckley and Leverett[1] in 1942. Strangely enough, the use of essentially the same equations for counterflow under gravity forces did not appear until 1959, with equations derived independently by Gilmour[5] and by Sheldon, Zondeck, and Cardwell[17].

Special functions have been devised to describe such physical properties and gas-liquid phase behavior of hydrocarbon mixtures as are pertinent to calculating reservoir performance. Such functions are, for instance, formation volume factors and solubilities referred to "stock tank" oil volumes, and a special type of "equilibrium constant." The determination of these functions, and other fluid properties such as gas and oil viscosities, under phase equilibrium conditions and reservoir temperature and pressures, has become standard laboratory practice since material balance and fluid flow equations were first applied to oil production.

It would be impossible to describe the hundreds of repressuring projects which have been in operation since gas injection was first started. Instead a few examples which show good engineering practice and which lend themselves to analysis will be described briefly. Questions and problems on these will be found in later chapters and at the end of the monograph, Appendix A.

1.1 The Jones Sand Pool[4]

One of the older well-engineered projects is the Jones Sand Pool in Arkansas discovered in September, 1937. It is an anticlinical trap covering a pear-shaped area of 4000 acres, 4 miles long and 1½ miles wide, with the dome about at the centroid at $-$ 7250 feet, the aquifer boundary at $-$ 7370 feet, and the gas oil contact at $-$ 7270 feet. The volume of productive oil zone was estimated at 150,000 acre-feet and the volume of the initial gas zone at 4000 acre-feet. Initially the reservoir pressure was 3520 psia at $-$ 7300 feet, temperature 198°F, gas solubility 765 SCF/STB and formation volume factor 1.45 RB/STB. In the pressure range 300 to 3520 psia, the latter two quantities are given by

$$R_s = 0.1875p + 105$$

and

$$B_O = 0.00008p + 1.168$$

The equilibrium gas formation volume factor is approximated by

$$B_g^{-1} = 0.348p - 22, SCF/RB$$

The average porosity is 0.202 and the average permeability 400 md, with considerable variation, including some dispersed shale. The initial stock tank oil in place was estimated at 100 million barrels by material balance, after enough time had elapsed to make proper allowance for water encroachment. The encroachment rate seemed to be proportional to the decline in reservoir pressure below the original, that is, 80 bbl per month per psi decline.

Some field data are tabulated in Table 1. Gas injection into the cap was started in 1941, when the average pressure had declined to 1542 psia. During the previous depletion stage, the pressure decline was roughly proportional to the oil production, 1 psi per 9500 bbl. From July, 1941, to March, 1955, most of the gas was reinjected, and the pressure nearly stabilized at slightly under 1500 psia. Water injection in down dip locations began on a small scale in July, 1944, and became intensified two years later. Gas injection was sharply curtailed about March, 1955, and wells with high gas-oil ratios shut in. Gas injection was abandoned at the end of 1957, but water injection continued. The average reservoir pressure rose from a low of 1200 psia in 1954 to 1800 in 1962. All wells which remained on production after the end of 1957 had gas-oil ratios approximating the solution values. In January, 1962, the recovery was 65 per cent of the original oil in place, with 70 per cent expected at ultimate abandonment.

Table 1. Jones Sand Pool Field Data

		Gas-Oil Ratio, SCF/STB			Water, Mbbl			
end of month	reservoir pressure, psia	prod. for month	net for month	net cumul.	prod.	net prod., $W_p - W_{inj}$	oil prod., MM STB	W_e MM bbl
9-37		750	750	750	0	0	0	0.
7-38	3153	950	950	917	0	0	2.82	0.1
7-39	2533	1350	1350	1054	0	0	8.75	0.9
7-40	1978	2150	2150	1331	0	0	14.00	2.1
7-41	1542	1834	955	1512	5	5	19.01	3.9
7-42	1510	1780	141	1256	94	94	23.72	5.8
7-43	1490	2238	97	1082	306	306	27.92	7.8
7-45	1498	3312	-36	868	1110	963	34.99	11.8
3-47	1471	3468	229	763	1163	621	40.43	15.2
3-48	1454	3666	397	759	2367	- 46	43.70	17.2
3-49	1429	4648	755	737	3925	-804	46.90	19.3
3-50	1432	4393	638	726	5398	-1749	49.41	21.3
3-51	1335	4600	770	723	4037	-3328	51.84	
3-53	1335	6000	1100	718	5091	-7950	55.36	
3-55	1405	3500	1150	731	6166	-13476	58.17	
3-57	1440	1200	1000	742	7646	-21012	60.51	
3-59	1750	352	352	736	10172	-26797	62.64	
3-61	1940	474	474	725	13215	-30291	64.87	

1.2 Mile Six Pool[1][2]

The Mile Six Pool in Peru is a classical text book example of the benefits that can be obtained by gas injection where conditions are such that advantage may be taken of gravity forces. The pool was discovered in 1927 and opened in September, 1933, and gas returned to the cap almost from the beginning. The pool is a monocline, with the surface area roughly resembling a quadrant of an ellipse. The dip is about 17°, with the cap near the "focus," edge water at the curved "perimeter," and a fault along the "semi-major axis." An initial gas cap occupied about 8 per cent of the total sand volume of 54,000 acre-feet. The average absolute permeability is of the order of 1000 md, and the average hydrocarbon porosity 0.163. The original equilibrium oil and gas viscosities have been estimated as 1.32 and 0.0134 cp, densities 0.78 and 0.08 gm/cc, and formation volume factors 1.25 RB/STB and 0.0025 RB/SCF. The previous figures give initial oil in place as 50 million STB.

The initial reservoir pressure of about 930 psia dropped slightly the first year to about 850 psia, where it remained until 1946 when a blowout caused a drop to 750 psia. The field producing gas-oil ratio averaged 400 SCF/STB (probably the initial solubility) from 1934 through 1938, 600 from 1939 through 1941, then rose to a maximum of about 2300 by 1945, after which it declined to about 1200. The rise was apparently due to keeping some wells producing after passage of the oil-gas "front," since 85 per cent of the ultimate production had been reached and an irregular oil-gas "front" was near the base of the reservoir. Later some wells were reconditioned and selectively perforated so that

year	gas-oil ratio (av. for year)		reservoir pressure, psia	cum. oil prod. (end of year), MM STB
	Prod.	Inj.		
1933			930	0.
1934	300	300	840	3.7
1935	300	400	850	7.8
1936	300	600	840	12.5
1937	400	700	840	17.5
1938	400	550	860	20.7
1939	700	800	860	22.5
1940	600	800	870	23.7
1941	600	700	860	24.8
1942	1100	1200	860	25.7
1943	1900	1800	850	26.5
1944	2300	2300	830	27.1
1945	2400	2500	825	27.5
1946	1200	1400	760	28.0
Ultimate				30.0(?)

Table 2. Mile Six Pool Production Data

production was mostly from oil-saturated regions. Some water entered edge wells shortly after the blowout, but retreated after the wild well had been capped. The gas-oil and water-oil interfaces seemed to remain quite horizontal, with almost no water encroachment. The table gives some smoothed production data[2].

1.3 Frio Sand Reservoir[4]

The multiple producing-zone Frio sands are in south Texas. A brief summary will be given of a gas injection project in one of these zones, the "D-5 reservoir." The structure is a large asymmetrical nose plunging gently (about 0.5°) to the northeast. The dip toward the northwest (the main oil producing flank) is 1.5°. The reservoir is bounded on the latter flank by an oil-water contact, on the north and south by the sand lens grading into shale, and on the east by a steep fold or fault parallel to the axis of the structure (see Figure 1.1). The average depth is 3750 feet. The oil reservoir covered 1870 acres with an average thickness of 10.1 feet, and the original gas cap 400 acres with an average net thickness of 10.6 feet giving an average ratio of gas cap to oil reservoir of 0.225.

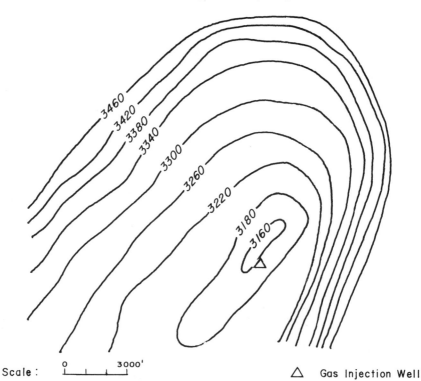

Scale : 0 ⊢—⊥—⊥—⊣ 3000'   △   Gas Injection Well

Figure 1.1. Frio sand reservoir.

Other pertinent data are as follows: ϕ = 0.23; k = 194 md (av); S_w = 0.24 assumed uniform and constant; p_i = 1590 psia; B_o = 1.06 + .000073 p; B_{oi} = 1.1768; R_s = 35 + .178ρ; R_{si} = 319 SCF/STB; Temp. = 154° F = 614°R; Z = 0.995 - 0.00015 ρ; Z_i = 0.757; B_g = 3.07Z/p; B_{gi} = 0.00146; Then N = 21.74 MMSTB; G_{soln} = 6.96 MMMSCF; G_{cap} = 3.94 MMMSCF. Sat. oil visc., μ_o = 0.86 − .00014 p; Sat. gas visc. μ_g = .0115 + .000003 p; More simply viscosity ratio μ_o / μ_g = 70 − 0.02 p.

Production and injection data are given in Table 3.

end of year	cum oil prod. MMSTB	cum. gas prod. MMMSCF	cum. Gas Inj. MMMSCF	cum. water prod. MMBbl
1940	0.148	0.103	0.	0.
1941	.425	.239	.304	0.007
1942	.893	.482	.974	.010
1943	1.469	.962	2.218	.013
1944	2.126	2.037	2.874	.021
1946	3.193	4.852	4.548	.040
1948	4.353	8.434	6.343	0.72
1950	4.987	11.077	8.207	.114
1952	5.572	14.09	10.32	.182
1954	6.018	16.56	12.65	.251
1956	6.419	18.30	13.11	.307
1958	6.737	19.65	13.51	.363
1960	6.969	20.73	13.66	.424

Table 3. Production and Injection Data

While the dip was not enough to greatly affect the displacement efficiency in the individual pores, it apparently was enough to give a fairly well defined "front" ahead of the injected gas, that is, enough to "smooth out" formation heterogeneities to a considerable extent.

1.4 Brookhaven Field[4]

The Brookhaven field in Mississippi is an example of an initially undersaturated field with wide variations in saturation, pressures and amounts of gas dissolved. It is an anticlinal type structure with several faults which give partial discontinuities. There was a weak water drive with lack of complete communicability. Initial conditions may be summarized: Pressure 4500-4700 psia, gas dissolved 100-500 SCF/STB, porosity 0.256 (av.), water saturation 0.45, oil gravity 26°-40° API, average formation volume factor 1.25, sand volume 207,000 acre-feet, average thickness 30 feet, depth ca. 10,000 feet, temperature 250°F, dip 2°. The previous figures give 180 MMSTB of oil initially in place. The analysis will be confined to "Block IV," with a sand volume of 55,500 acre-feet, giving 48 MMSTB of oil initially in place.

The field was opened in early 1946 and gas injection started in early 1948 after production of 2.4 MMSTB from Block IV and a pressure decline of 2000 psia. Gas was injected to maintain an average pressure of 2850 psia behind the front. At this pressure and the reservoir temperature, for the saturated oil, B_o = 1.4, R_s = 647, μ_o = 0.54 cp, and, for the equilibrium gas, μ_g = 0.019 cp, B_g = 0.00121 RB/SCF. For the undersaturated oil R_s = 433 (av), B_o = 1.30. Relative oil permeability and gas-oil permeability ratio may be tabulated (S_w = 0.45):

S_g:	0	.05	.10	.15	.20	.25	.30
k_{ro}:	1	.43	.25	.13	.07	.03	.01
k_g/k_o:	0	.01	.075	.33	1.5	5.5	40.

From the start of injection to October, 1952, an additional 2.41 MMSTB of oil were produced ahead of the gas front and 3.04 MMSTB from behind the front. In the same period 14.0 MMMSCF of gas had been injected, and 11.2 MMMSCF produced. It was estimated that, in October, 1952, 24,600 acre-feet of sand was behind the gas front, and that 0.7 MMB of water had invaded. A "coverage" factor of 0.55 was estimated, as illustrated by a problem in Appendix A. In general, wells were not shut in after passage of the gas front, but continued on production to a gas-oil ratio of 50,000 SCF/STB.

1.5 Miscible Drives

It was recognized, almost from the beginning of the industry, that a certain "residual saturation" of oil would be left after a drive with an immiscible fluid such as gas, air, or water. It was also known, however, from core testing, that oil could be completely removed from porous media by solvents such as naphtha. The use of solvents for oil recovery in the field was considered impractical because, in the subsequent attempts to recover the solvent by an immiscible drive, there would remain this unrecoverable or "residual" saturation of solvent. Thus unrecoverable solvent would be substituted for unrecoverable oil, necessitating a very cheap solvent. Such a solvent has never been discovered.

It was not until about 1950 that petroleum engineers really became aware of the quantitative relations involved in "miscible displacement" in porous media, although the concepts were well known to chemical engineers much earlier. It soon became apparent, from the mechanisms and quantitative relations involved, that complete substitution of solvent for oil or filling the reservoir with solvent were not necessary in solvent or "miscible" methods of oil recovery. The first promising method using only a limited amount of solvent was the "propane slug method" suggested by Atlantic Refining Company (now Atlantic Richfield)[8].

If the solvent (e.g. propane) were continously injected and moving through the formation, it would form a solution with the oil, and there would be a transition zone of continuously varying ratio of oil to solvent. In the forward part of the zone the hydrocarbon phase is essentially all reservoir oil. The ratio of solvent to oil increases through the zone, in a direction opposite to the flow direction, the composition being essentially pure solvent at the upstream part of the transition zone. The volume of this transition zone continues to increase as it moves through the formation. The quantitative relations are discussed in Chapter 4.

If this transition zone were "pushed" until it was produced at wells, all the oil from the reservoir would, theoretically, be produced, since the oil would have moved ahead of the transition zone. The problem, then, was to find a cheap or cheaply recoverable "pusher." Driving the solvent with an immiscible phase such as water or air (or low pressure gas) would, as mentioned above, leave a residual solvent saturation, meaning a prohibitive outlay subsequent loss of solvent. The Atlantic process takes advantage of the fact that, at reasonably high pressures, a continuous phase may be formed in which the composition varies from liquid propane to natural gas. That is, it is possible to form a total single phase hydrocarbon transition zone varying from reservoir oil to natural gas, the middle part being largely propane of LPG (liquified petroleum gas). The Atlantic process mentioned above has been called the "propane slug process" because a limited amount ("slug") of solvent is injected and followed by gas, the pressure being enough to maintain a single phase. The gas "pusher" may in turn be pushed by something cheaper such as air (or air followed by or alternated with water) but, since an expanding gas-air transition zone would also be formed, the natural gas zone must be wide enough so that miscibility is not lost due to subsequent overlapping of the air-gas and gas-propane transition zones, as they grow continuously in volume.

Instead of injecting a solvent "slug," it is possible in some cases to form the miscible drive "in situ." One possibility involves the "enriched gas drive" in which the injected gas contains enogh light condensible hydrocarbons (e.g. propane and butanes) to form the miscible zone as a result of the reservoir pressure and contacts with the reservoir oil. This has also been called a "condensing gas drive." Another possibility is the "high pressure gas drive" or "vaporizing gas drive" in which the more volatile components of the reservoir oil enter the gas phase so that, by repeated contacts, a miscible zone is formed. The theory of these "in situ" methods of forming a miscible drive is discussed in Chapter 4.

1.6 Miscible Slug Injection in a Pattern — Parks Field[1][1]

Parks field in Midland County, Texas, is a double anticline of Pennsylvanian limestone bounded by faults on one side and thinning out on the rest of the boundary. The oil-water contact is at the thin part, and material balance calculations showed little water encroachment. The porosity is fossiliferrous, averaging 0.068, the permeability averages 2.5 md, the dip about 1°, the net pay thickness 20 feet, the total thickness 61 feet, and the depth about 10,000 feet. The productive area is 6400 acres and the water saturation 0.33. From this information and an initial formation volume factor of 1.93, the amount of oil originally in place is calculated as 23 MMSTB. The field was initially undersaturated at 4560 psia, the bubble point of the original oil being 3500 psia. During primary production, from the opening of the field in 1951 to July, 1957, 3 MMSTB of 45° oil were produced, with the pressure dropping to 1900 psia, and roughly 15 MMMSCF of gas being produced. At 1900 psia, the oil formation volume factor was 1.50 and the viscosity 0.384 cp. The viscosity was 0.22 cp originally at reservoir conditions, that is, 4560 psia and 174°F.

Because of low permeability and dip, secondary methods required a pattern, and this was selected so that 15 injection wells were near the field boundaries and 8 in the more interior area. From July, 1957, to July, 1958, 1.54 MM barrels of propane were injected (4 per cent of the hydrocarbon pore volume) and this was followed by gas, maintaining the reservoir pressure, which was above that required for miscibility. Water was injected later to improve sweep efficiency. Up to July, 1960, 6.87 MMMSCF of gas had been injected, and 2.56 MMB of water had been injected in a pilot water flood. Cumulative production was 4.2 MMSTB of oil, 24.8 MMMSCF of gas, and 0.11 MMB of water. From propane breakthrough times, a 50 per cent vertical sweep efficiency to breakthrough was indicated, about four times that given by a Stiles type calculation. The vertical permeability profile assumed was not given, although there was mention of three correlatable zones of porosity. Potentiometric models were used to obtain areal sweep efficiencies, but these were not given. A map indicates about 70 per cent to breakthrough. All sweep efficiencies were calculated assuming unit mobility ratio, in anticipation of water injection behind or with the gas. Considering that many wells would be shut in after propane breakthrough to eliminate the cusp at the front, and considering production after breakthrough in other wells, an ultimate recovery of at least 55 per cent was expected, as compared with 41 per cent for a water flood.

The fact that production increased and gas-oil ratios decreased long before propane breakthrough indicated bank formation. Failure of the Stiles type calculation, and the fact that the propane slug was fairly well distributed vertically when injected, indicate good vertical communication as well as change in permeability profile with horizontal distance. No attempt was made to calculate "overriding" or deformation of the slug due to density differences.

1.7 Propane Slug Oriented by Gravity — Baskinton Field[9]

A propane slug drive which was aided by gravity segregation was carried out in Baskinton Field, Louisiana. This is a stratigraphic trap like an elliptical lens, major axis about twice the minor axis, about 40 feet maximum thickness and thinning out toward the circumference. The area is 255 acres, with a $2°$ dip southward along the minor axis. The thickest part is about two-thirds of the way down dip so that the pinching out is fastest in the lower (southern) third. Primary production was from a few (probably five) wells spaced roughly around the 20 foot isopach. From the opening, about July 1, 1956, to January 1, 1962, production included 0.525 MMSTB of oil, 0.275 MMMSCF of gas, and 0.006 MMB of water, with the pressure declining from the original 1800 psia to 1650 psia. The initial oil in place was estimated at 4.9 MMSTB, with a formation volume factor of 1.372, and saturated with a gas solubility of 600 SCF/STB. A material balance indicated a small water drive at the lower edge of the field. The porosity was given as 0.29, water saturation 0.30, and permeability 400 md.

Propane injection was started in April, 1962, and 0.21 MMB were injected by September 4. This was followed by gas, the pressure and temperature ($162°F$) being sufficient to give a continuous phase, oil to propane to gas. The injection well was near the north-central pinchout where the formation is fairly thin, about seven feet. Because of the thickening down dip, it was decided to locate an observation well down dip, in a 30-foot section about half way between the injection well and down dip producers, to determine the orientation of the slug. This well was perforated near the top and bottom and at the midpoint of the sand, and withdrawls were slow so as to avoid major fluid movements.

The rapid appearance of nearly pure propane and the persistence of samples that were predominantly propane at only the upper perforations led to the conclusion that the slug was a nearly horizontal layer. The reservoir withdrawal rate was adjusted so that the "critical" rate to prevent vertical viscous fingering was not exceeded. The slug was considered thick enough so that miscibility would not be lost due to growth of the transition zones.

1.8 High Pressure Miscible Drive — Block 31 Field[19]

The classical example of a high pressure (vaporizing) gas drive in which miscibility was obtained was University Block 31 Field in Texas, operated by Atlantic Refining Company (now Atlantic Richfield). The main oil producing section is the Middle Devonian, which is the part being miscibly swept. It is a 350-foot interval with permeable zones that can be correlated from well to well, with a maximum of 185 feet of net pay near thy crest of the structure, which is an anticline. No mention was made of an oil-water interface.

The field was discovered near the beginning of 1946 and, by June, 1949, had produced 3.7 MMSTB of $46°$ API oil by primary depletion. The pressure had dropped from an initial 4160 psia to 3665 psia. The saturation pressure was 2780 psia, with a formation volume factor of 1.69 and gas solubility 1300 SCF/STB, and viscosity 0.3 cp at the reservoir temperature of 139°F. The average porosity is 0.15, permeability 1 md, and water saturation 0.37. The initial oil in place was estimated at 275 MMSTB, the area of the unit being about 11 square miles.

Since it was estimated that a depletion drive would yield only 25 per cent of the oil in place, unitization with pressure maintenance was decided upon. Injection was mainly in the crest at first, but it was found that the reservoir pressure could not be maintained with economical withdrawal rates. Meanwhile it was also found that the average reservoir pressure had to be maintained above 3200 psia for a miscible drive, that is a pressure above 3500 at the gas-oil transition zones. Hence a nine-spot injection pattern over the unit was adopted (one injection well for three producing wells). The possibility of a miscible drive had been demonstrated by laboratory tests and by the phase diagram for the gas and oil. The theory is discussed in Chapter 4.

After some additions, the final compressor plant was designed to deliver 57 MMSCF per day at 4000 psi, of which 60 per cent was purchased. Injection wells were capable of taking 2 to 10 MMSCF/day, those at the crest taking more than those near the flanks. Production was balanced against intake in each pattern unit.

There were some leaks into non-pay zones, and these were traced by radioactive tracer techniques and largely remedied. Plugging of some injection wells presented a problem. This apparently was due to compressor lubricant in the gas, formation of iron sulfide from hydrogen sulfide, and oxidation of hydrogen sulfide to sulfur by oxygen in the gas. A gasoline-xylene-acid was found effective in restoring plugged wells.

In 1961[6] the field had produced about 55 MMSTB of oil, with an expected ultimate recovery of about 150 MMSTB. This figure would probably be revised upward by alternate injection of gas and water to increase sweep efficiency. Flue gas was later substituted for natural gas.

1.9 Enriched (Condensing) Gas Drive Project -- Seeligson Zone[18]

This zone may be roughly regarded as dipping eastward from a fault line at about 4°, and bounded by strand lines on the north, south, and the southern part of the east limits by strand lines, with an oil-water contact at the northern third of the east boundary. The productive area is 877 acres and average thickness 12 feet. The permeability averages 100 md, porosity 0.19, connate water saturation 0.31, and temperature 171°F. The oil was initially saturated at 3010 psia. The gas solubility was given as 640 SCF/STB and the formation volume factor as 1.37 at a saturation pressure of 2380 psia. The stock tank oil had a gravity of 44° API. A small gas cap was present.

Before gas injection the pressure had dropped from 3010 to 2400 psi with the production of 0.75 MMSTB of the 7.4 MMSTB originally in place. Material balance calculations indicated little water encroachment. Primary (depletion) production was estimated at 22.3 per cent.

Injection of a 50:50 (moles or gaseous volumes) mixture of propane and residue gas was started up structure in March, 1957. The average rate was 3.575 MMSCF per day into two wells. Thy southernmost injection well and two southernmost producing wells were closed later, after an early breakthrough. The propane mixture was continued until September, 1960, after 0.20 hydrocarbon pore volumes had been injected. At that time a similar ethane-dry gas mixture was started and continued until February, 1963, when 0.37 hydrocarbon pore volume of enriched gas had been injected. Appearance of propane and ethane at the various rows of wells indicated a volumetric breakthrough efficiency of about 21 per cent. This is a combination of horizontal sweep, largely governed by mobility ratio, and vertical sweep, governed by permeability variations and gravity effects. Tests made on a down dip well indicated permeability distribution as a major factor, the top of the formation being the more permeable. For the purpose of problems in a later chapter, one might assume, as an oversimplification, 20 feet of 200 md sand overlying 20 feet of 75 md sand. Portions which had been well contacted by the enriched gas were found to be swept clean of oil.

From March, 1963, to February, 1964, water and propane-enriched gas were injected simultaneously at a rate of 1070 barrels of water plus 2.23 MMSCF of gas per day. This did not match withdrawal rates at reservoir volumes, and the reservoir pressure began to decline. During the water injection period, gas production exceeded gas injection. Attempts to increase injectivities by using alternate slugs of water and enriched gas (2770 bbl of water per day for 3 months and 2.95 MMSCF of gas per day for 5 months) beginning February, 1964, were unsuccessful. All injection was stopped in June, 1965.

At the start of gas injection, oil production was increased from about 300 to about 500 STB per day, with the gas injection sufficient to build up the reservoir pressure to about 2700 psia. From July, 1958, to June, 1965, when all injection stopped, the oil production was about 1000 STB/D, and the gas injection maintained the reservoir pressure at about 2700 psia until water was injected with the gas. From June, 1965, the termination of injection, to July, 1967, the daily oil production averaged about 300 STB. During the water injection period, the reservoir pressure dropped from 2700 to 2200 psi, then dropped more rapidly, reaching 1400 psi by July, 1967.

Average producing gas-oil ratios remained in the neighborhood of 1000 SCF/STB until July, 1960, apparently by shutting in wells behind the front. From July, 1960, to July, 1967, the producing gas-oil ratio rose regularly to 6000, with a peak of 9000 in the last half of 1965, following the stop of injection.

Large volumes of oil were produced after breakthrough of the enriched gas front. For instance, in January, 1962, "75 per cent of the reservoir volume was behind the breakthrough front" (apparently considered as a vertical surface penetrating the formation) with only 30.5 per cent recovery of the original oil in place. But 50 per cent of the oil had been recovered by July, 1967, with 84 per cent of the reservoir behind the said "front."

The 50 per cent recovery may be compared with 22.3 per cent estimated for a depletion operation and 42 to 48 per cent estimated for a water drive, from data on similar reservoirs in the area. An additional one per cent was expected from subsequent pressure blowdown and 3 to 4 per cent from water flooding the northeast segment not swept by the gas injection project. Details on recovery of propane and ethane were not given, and apparently, all injected gas was enriched.

1.N Symbols and Units

The letter symbols used in the mathematical expressions throughout this monograph, are almost entirely those recommended by AIME. In a few cases it became necessary to use some special secondary symbols such as serifs, tildes, subscripts and superscripts so as to avoid ambiguity.

By the same token, field units, as opposed to Darcy units, are generally applicable to the many formulas given for quick calculations. Table 4 shows the equivalence between two systems of units and also indicates that the complete carry over from the Darcy to the field system is easily accomplished by simply expressing the permeability in perms, that is, $6.33k_{darcy}$.

Table 5 gives some useful conversion factors that the petroleum engineer frequently needs.

Table 4. Equivalents, Darcy, and Field Units		
Basic parameters	Darcy units	Field units
q	cc/sec	ft³/day
μ	cp	cp
k	darcy	$6.33k_{darcy}$
p	atm	psi
r	cm	ft
t	sec	day
c	atm⁻¹	psi⁻¹

Table 5. Conversion Factors							
158984.	cc	=	1 bbl	43560.	ft²	=	1 acre
5.615	ft³	=	1 bbl	62.43	lb/ft	=	1 gm/cc
6.290	bbl	=	1 m³	379.48	SCF	=	1 lb mole
7758.	bbl	=	1 acre-ft	3.968	BTU	=	1 Kcal
35.315	SCF	=	1 m³	50.	ton/yr	~	1 bbl/day
30.48	cm	=	1 ft	10.	m³/yr	~	1 SCF/day

References

1 Buckley, S.E., and Leverett, M.C. "Mechanism of Fluid Displacement in Sands." *Trans. AIME* (1942) 146, 125.
2 Craft, B.C., and Hawkins, M.F. *Applied Petroleum Reservoir Engineering,* Prentice-Hall, 1959.
3 D'Arcy, H. *Les Fontaines publique de la ville de Dijon,* Victor Dalmont, Paris, 1856.
4 Field Case Histories: Oil Reservoirs. Petroleum Transactions Reprint Series, *Soc. Pet. Engs. of AIME,* Dallas, 1962.
5 Gilmour, H. "Gas and Oil Segregation in Oilfield Reservoirs." *J. Inst. Pet.* (Dec., 1960) Vol. 46, No. 444, 1.
6 Herbeck, E.F., and Blanton, J.R. "Ten Years of Miscible Displacement in Block 31 Field." *J. Pet. Tech.* (June, 1961) 543-549.
7 Herold, S.C. *Analytical Principles of the Production of Oil, Gas, and Water from Wells,* Stanford University Press, 1928.
8 Koch, H.A. Jr., and Slobod, R.L. "Miscible Slug Process." *Trans. AIME* (1957) 210, 40.
9 Laue, L.C., Teubner, W.C., and Campbell, A.W. "Gravity Segregation in a Propane Slug Project – Baskinton Field, Louisiana." *J. Pet. Tech.* (June, 1965) 661-3.
10 Leverett, M.C. "Flow of Oil-Water Mixtures Through Unconsolidated Sands." *Trans. AIME* (1939) 132, 149.
11 Marrs, D.G. "Field Results of Miscible Displacement Program Using Liquid Propane Driven by Gas, Parks Field Unit, Midland County, Texas." *J. Pet. Tech.* (April 1961) 327-332.
12 Moyar, R.E. "A Review of Peruvian Operations – Mile Six Pool." *O. & G. Jour.* (Dec. 27, 1947) 46,251.
13 Muskat, M. *Flow of Homogeneous Fluids in Porous Media,* J.W. Edwards, Inc., Ann Arbour, Mich., 1938.
14 Muskat, M. *Physical Principles of Oil Production,* New York. McGraw-Hill, Inc., 1949.
15 Schilthuis, R.J. "Active Oil and Reservoir Energy." *Trans. AIME* (1936) 118, 33.
16 *Secondary Recovery of Oil in the United States,* 2nd. Edition, American Petroleum Institute, New York, 1950.
17 Sheldon, J.W., Zondek, B., and Cardwell, W.T., Jr. "One-Dimensional Incompressible, Non-Capillary, Two-Phase Fluid Flow in a Porous Medium." *Trans. AIME* (1959) 216, 290.
18 Walker, J.W., and Turner, J.L. "Performance of Seeligson Zone 20B-B07 Enriched Gas Drive Project." *J. Pet. Tech.* (April, 1968) 369, 373.
19 Whorton, L.P., and Kieschnick, W.L. "Oil Recovery by High Pressure Gas Injection." *O. & G. Jour.* (April 6, 1950) 48, 78.
20 Wyckoff, R.D., and Botset, H.G. "The Flow of Gas-Liquid Mixtures Through Unconsolidated Sands." *Physics* (1936) 7, 325.

2. Material-Balance Methods

2.0 Introduction

The most important development in petroleum engineering during the past decade has been the extensive use of reservoir simulators for simulating complex problems which could not have been solved by conventional material balance methods. The material balance approach, however, constitutes a valuable tool in gaining an understanding of reservoir mechanics and is particularly useful in providing the engineer with quick and fairly accurate answers at almost negligible computer cost. Moreover, conventional material balance techniques are necessary for early reservoir evaluation in preparation, perhaps, for the more detailed simulation study. This early evaluation complements the model study and is fundamental in the success of the model.

As generally used for pressure depletion below the bubble point, the overall balances depend on constancy of total pore volume, uniform pressure with phase equilibrium temperature and fluid composition and, of course, correct values of volumetric equilibrium fluid properties at reservoir temperature. Some discrepancy can arise because of a difference in conditions of gas liberation in the reservoir as compared with the laboratory. Nevertheless, if the previous conditions are met, the balances do not depend on saturation distributions or fluid movements within the reservoir confines.

It is this latter feature that makes possible a unified approach in analysis and prediction. The presence of regions invaded by water or the expanding gas cap (or injected gas), or fluid distributions due to gravity segregation, conformance, coverage, etc., do not invalidate the overall balance equations with the necessary terms.

This has led to their usefulness in the analysis of the performance of combination drive reservoirs. Deviations from the assumptions given earlier do not seem to be really serious in most cases. The build-up methods of well pressure measurement, for instance, help to give a fair average for the formation.

This chapter presents a general approach to developing the material balance equations for the various displacement mechanisms encountered in petroleum reservoirs. Particular interest is given to combination mechanisms involving segregation drives with pressure depletion and also in repressuring. Both the integral and differential forms of the material balance equations are discussed emphasizing the convenience of each form in relation to the type of problem to be solved.

2.1 The Zero-Dimensional Equation

With the advent of sophisticated numerical techniques it is now quite common to simulate the multiphase flow of compressible or incompressible fluids in a single or multi-dimensional porous medium, by solving the pertinent system of partial differential equations. Before beginning a discussion of the material balance equations, it is instructive to put into perspective the relationship between these two types of approaches, the limitations and advantages of the latter.

Consider a porous medium saturated with oil, gas and immobile water. The controlling equations can be shown to be[6].

Oil phase:

$$\nabla \left\{ \frac{\lambda_o}{B_o} \; \nabla \; (p_o + \rho_o \, gz) \right\} = \phi \; \frac{\partial}{\partial t} \left(\frac{S_o}{B_o} \right) \qquad 1$$

Gas Phase:

$$\nabla \left\{ \left(\rho_g \lambda_g + \frac{R_s \lambda_o}{B_o} \right) \nabla \; \left(P_g + \rho_g \, gz \right) \right\}$$

$$= \phi \; \frac{\partial}{\partial t} \left(\rho_g S_g + \frac{R_s S_o}{B_o} \right) \qquad 2$$

Capillary Pressure:

$$P_c = P_g - P_o \qquad 3$$

Saturation:

$$S_{wi} + S_o + S_g = 1 \qquad\qquad 4$$

$$\lambda_o = \frac{k_o}{\mu_o} \; ; \; \lambda_g = \frac{k_g}{\mu_g}$$

and z is positive upward.

Equation 1, in the absence of gravitational ($\rho g z$ = o) and capillary forces (P_c = o), is integrated over the entire reservoir volume V_r. We obtain

$$\int_{V_r} \nabla \; \left(\frac{\lambda_o}{B_o} \nabla \; p \right) \, dV_r = \phi \int_{V_r} \frac{\partial}{\partial t} \left(\frac{S_o}{B_o} \right) dV_r \qquad\qquad 5$$

which, by the divergence theorem, becomes

$$\int_{A_r} \frac{\lambda_o}{B_o} \frac{\partial p}{\partial n} \; dA_r = \phi \int_{V_r} \frac{\partial}{\partial t} \left(\frac{S_o}{B_o} \right) dV_r \qquad\qquad 6$$

where $\dfrac{\partial}{\partial n}$ denotes differentiation along the normal to the reservoir area A_r of the volume V_r and directed away from the interior of A_r. Assuming average values over the integral, we can write

$$\frac{\lambda_o A_r}{B_o} \frac{\partial p}{\partial n} = \phi \; V_r \frac{\partial}{\partial t} \left(\frac{S_o}{B_o} \right) \qquad\qquad 7$$

The left hand side of this equation is easily recognized to be Darcy's equation for the oil phase. Thus we may write

$$q_o = - \phi \; V_r \frac{\partial}{\partial t} \left(\frac{S_o}{B_o} \right) \qquad\qquad 8$$

Similarly, **Equation 2** can be integrated to yield

$$q_g = - \phi \; V_r \frac{\partial}{\partial t} \left(\rho_g S_g + \frac{R_s S_o}{B_o} \right) \qquad\qquad 9$$

The producing gas-oil ratio of the reservoir at surface conditions is defined as

$$R = \frac{q_g\,B_o}{q_o\,B_g} + R_s = \lambda\,\frac{B_o}{B_g} + R_s \qquad\qquad 10$$

where

$$\lambda = \lambda_g/\lambda_o \qquad\qquad 11$$

Substituting the right hand sides of **Equations 8** and **9** into **Equation 10** and changing the time variable of differentiation to that of pressure, gives the following relationship

$$\frac{d\,S_o}{d\,p} = f_o \left\{ -\frac{S_g}{B_g}\frac{d\,B_g}{dp} + \frac{B_g\,S_o}{B_o}\frac{d\,R_s}{dp} + \frac{\lambda\,S_o}{B_o}\frac{d\,B_o}{dp} \right\} \qquad\qquad 12$$

where,

$$f_o = (1 + \lambda)^{-1} \qquad\qquad 13$$

Equation 12 is known as Muskat's material balance equation (abbreviated MBE throughout this chapter) or the semi-steady state depletion equation since it represents an integral average of the true flow equations where saturation and pressure gradients are considered uniform. In other words, no space or time variation of the reservoir parameters is considered and for this reason, **Equation 12** is also referred to as the zero-dimensional approximation of the depletion problem.

It is obvious that the solution of the ordinary differential **Equation 12** is several orders of magnitude simpler than that of the system of partial differential **Equations 1** through **4**. Most important, field observations have been in remarkably good agreement with the material balance calculations. This may be expected for two reasons. First, most of the pressure and saturation gradients in radial flow occupy only a small fraction of the total reservoir volume. Secondly, from a physical point of view, semi-steady state conditions prevail in most reservoirs after small production times, $t \sim \dfrac{r_b{}^2}{3\eta}$. r_b refers to the equivalent reservoir radius and η is the hydraulic diffusivity constant, all expressed in field units.

Consider, for example, a typical bubble point reservoir with a five mile radial extension and a hydraulic diffusivity constant of $4 \cdot 10^6$ ft.^{2}/day, then

$$t \sim \frac{(5\cdot5280)^2}{3\cdot4\cdot10^6} = 58 \text{ days}$$

That is, the reservoir attains semi-steady state conditions in less than two months. In the case of gas reservoirs, however, η may be as much as ten times smaller than that for oil reservoirs, thereby extending appreciably the transient period of the reservoir and consequently limiting the material balance approach during the early life of the field. This does not imply that the MBE is not valid during transient flow, but that its representation of the dynamics of the reservoir is impaired.

From a theoretical point of view, the zero-dimensional solution, **Equation 12**, for the depletion model problem tends to yield lower gas-oil ratios and higher oil recoveries as compared to those obtained from the beta model, **Equations 1** through **4**. Also, when compared to the compositional model in which both dynamic and stoichiometric variations of the reservoir fluids are considered, the material balance results tend to lie below those of the beta and the compositional model, as shown in Figure 2.1 for a linear porous medium. Obviously, for the more realistic radial system the results of the three models should be closer.

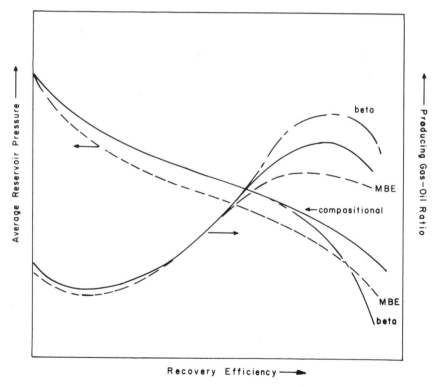

Figure 2.1. Qualitative performance curves for a depletion drive linear reservoir (after Roebuck et al[9]).

In a subsequent section (Section 2.5), **Equation 12** will be derived exclusively from the conservation point of view, thereby emphasizing its relationship to the other forms of the material balance equation.

2.2 Integral Material Balance Equations

From a historical point of view, the integral form of the mass conservation equations for the oil, gas and water phases and the volume conservation of the reservoir pore space, constitute the earliest models used to simulate reservoir performance problems.

These models have always been popular in engineering work since their algebraic simplicity not only makes them accessible to desk computations, but allows the engineer to literally build the model most suited for his particular problem.

While there is no limit to the degree of complexity that can be incorporated into these models, too much sophistication would seem to defeat the purpose, since dynamic simulators which make use of the beta model, **Equations 1** and **4** in section 2.1, are now readily available. In view of this, the derivations and discussions presented in this chapter will be limited to the basic assumption of uniform physical properties of the reservoir fluid and rock properties.

In general, the MBE is simply a linear relation among three basic variables and may be stated as:

$$\{Expansion\} - \{Withdrawals\} + \{Influx\} = 0 \qquad\qquad 1$$

The mathematical relationship among these variables is then readily established. In addition, the hydrocarbon reservoir (oil leg plus gas cap) is treated as a well-stirred tank model in thermodynamic equilibrium. Symbolically the model is shown in Figure 2.2.

Note that in as much as the MBE does not depend on uniform saturations, the "well stirred tank" analogy is not completely correct. However, the MBE does depend on uniformity of pressure and thermodynamic equilibrium.

Consider a saturated oil reservoir with an initial gas cap and an active aquifer. The gas-oil interface remains stationary and gas injected into the cap is assumed to disseminate freely throughout the oil leg, the only region limited to fluid withdrawals. It is further assumed that there is no oil saturation in the gas cap or in the aquifer and the residual water saturation is constant and equal throughout the hydrocarbon pore space. Likewise, the free gas saturation in the oil leg is initially zero and all fluids are considered incompressible.

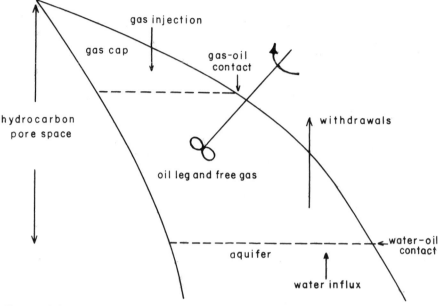

Figure 2.2. Well-stirred tank model of reservoir.

A material balance of the above system based on **Equation 1**, is as follows:

$\{System\ Expansion\}$

$$= \left\{ \begin{array}{c} Oil\ and \\ associated\ gas \end{array} \right\} + \left\{ \begin{array}{c} Gas\ cap \\ gas \end{array} \right\} + \left\{ \begin{array}{c} Residual \\ Water \end{array} \right\} - \left\{ \begin{array}{c} Pore \\ Volume \end{array} \right\}$$

$$= N\,(B_t - B_{ti}) + N\,m\,B_{ti} \left(\frac{B_g - B_{gi}}{B_{gi}} \right) \qquad\qquad 2$$

$$+ \frac{N\,(1+m)\,B_{ti}\,S_{wi}}{1 - S_{wi}} \left(\frac{B_{tw} - B_{twi}}{B_{twi}} \right) + \frac{N\,B_{ti}}{1 - S_{wi}}\,(1+m)\,c_f\,\Delta p \qquad 3$$

$$\{Withdrawals\} = \{Oil\} + \left\{ \begin{array}{c} Associated \\ plus\ free\ gas \end{array} \right\} + \{Water\}$$

$$= N_p B_o + N_p B_g\,(R_p - R_s) + W_p B_w \qquad\qquad 4$$

$\{Influx\} = \{Injected\ Gas\} + \{Encroached\ or\ Injected\ Water\}$

$$= G_i B_g + W_e \qquad\qquad 5$$

It should be noted in this development that expansion is considered positive whereas compression is negative; hence, the negative sign assigned to the shrinkage of the pore volume in **Equation 2**. We also used the following definition of the formation compressibility, namely

$$C_f \approx - \frac{1}{V_{pi}} \frac{\Delta V_p}{\Delta p} \qquad \qquad 6$$

where V_p is the reservoir pore volume and the subscript i denotes initial.

Similarly, the two phase formation volume factor B_t, has been introduced for the sake of compactness. By definition

$$B_t = B_o + (R_{si} - R_s) B_g \qquad \qquad 7$$

When **Equations 3, 4** and **5** are substituted into **Equation 1**, the following MBE is obtained:

$$N = \qquad \qquad 8$$

$$\frac{N_p \left\{ B_t + (R_p - R_{si}) B_g \right\} - W_e + W_p B_w}{B_t - B_{ti} + m B_{ti} \left(\dfrac{B_g - B_{gi}}{B_{gi}} \right) + \dfrac{(1 + m) B_{ti} S_{wi}}{1 - S_{wi}} \left(\dfrac{B_{tw} - B_{twi}}{B_{twi}} \right) + \dfrac{(1 + m) B_{ti} c_f \Delta p}{1 - S_{wi}}}$$

where the net cumulative gas-oil ratio, R_p, includes the injected gas; W_p is the net cumulative produced water and includes the injected water.

Equation 8 essentially involves three unknowns: (1) the original volume of oil N (2) the original volume of the gas cap Nm, and (3) the cumulative water encroachment into the reservoir during the production period, W_e. When, however, there are available independently established controlling data, such as the values of the initial oil in place or gas cap volume, the material balance equation provides a useful tool in estimating water intrusion or in predicting future reservoir behavior. On the other hand if all three of these parameters are unknowns, then the problem from a practical point of view, becomes indeterminate because of the difficulty of separating the variables. For any two unknowns the accuracy of the answers depends critically on the accuracy of the production data.

One of the main advantages of presenting the material balance in its generalized form, as above, lies in the fact that performance equations may be tailor-made for specific reservoirs by simply incorporating the relevant assumptions. In order to emphasize the value of this unified approach three well-known cases will be analyzed: (1) undersaturated reservoirs, (2) dry gas reservoirs and (3) depletion drive reservoirs.

Undersaturated Reservoirs

This type of reservoir is characterized by a deficiency of gas at the prevailing reservoir conditions. As a consequence, the reservoir does not contain a gas cap ($m=o$), the solution gas-oil ratio is constant and equal to the value at the bubble point ($R_s = R_{si}$) which implies that $B_t = B_o$ and $R_p = R_{si}$. Furthermore, since the oil above the bubble point pressure is virtually in an elastic state, the compressibilities of the formation and the connate water may now contribute appreciably to the energy of the system. With these assumptions, **Equation 8** reduces to

$$N \left\{ B_o - B_{oi} + \left(\frac{B_{oi}\ S_{wi}}{1 - S_{wi}} \right) \frac{B_w - B_{wi}}{B_{wi}} + \frac{B_{oi}\ c_f\ \Delta p}{1 - S_{wi}} \right\}$$

$$= N_p\ B_o - W_e + W_p\ B_w \qquad\qquad 9$$

This equation can be written in a more compact form, namely

$$N B_{oi}\ c_{oe}\ \Delta p = N_p\ B_o - W_e + W_p\ B_w \qquad\qquad 10$$

where the effective compressibility of the oil, c_{oe}, is defined as

$$c_{oe} = (S_{oi}\ c_o + S_{wi}\ c_w + c_f)/\ S_{oi} \qquad\qquad 11$$

Δp is the cumulative reservoir pressure decline.

Dry Gas Reservoirs

In the case of dry gas reservoirs the compressibilities of the formation and of the connate water can be considered negligible with respect to that of the gas phase so that the third and fourth terms in the denominator of **Equation 8** vanish. The process of further adapting the general material balance equation, **Equation 8**, to this specific problem, however, requires associating the individual terms therein with their physical meanings. For instance, the term involving the expansion of the oil and its associated gas, $N (B_t - B_{ti})$, becomes zero because the reservoir contains no oil. Likewise the gas cap expansion term, $Nm \dfrac{B_{ti}}{B_{gi}} (B_g - B_{gi})$, must now refer to the expansion of the entire gas reservoir and may be simply written as $G (B_g - B_{gi})$.

By the same token, the term in **Equation 8** that represents the cumulative production of oil and gas, namely, $N_p \left\{ B_t + (R_p - R_{si}) \right\}$ now refers in this context to cumulative free gas production, or, $G_p \, B_g$. After making these modifications and inserting the above assumptions, **Equation 8** becomes

$$G \, (B_g - B_{gi}) = G_p \, B_g - W_e + W_p \, B_w \qquad\qquad\qquad 12$$

Another useful form of **Equation 12** can be obtained by expressing explicitly the gas formation volume factor, B_g, in terms of the gas law, namely

$$B_g = 0.00504 \, \frac{TZ}{p}, RB/SCF \qquad\qquad\qquad 13$$

Under the assumption of isothermal reservoir conditions, **Equation 13** is substituted into **Equation 12** and the latter rearranged to give

$$\frac{p}{Z} \, (1 - \hat{W}_{\!q}) \;=\; \frac{p_i}{Z_i} \, \left(1 - \frac{G_p}{G} \right) \qquad\qquad\qquad 14$$

where

$$\hat{W}_{\!q} \;=\; \frac{W_{\!q}}{G \, B_{gi}}; \quad W_{\!q} \;=\; W_e - W_p \, B_w \qquad\qquad\qquad 15$$

T is the reservoir temperature and p_i/Z_i and B_{gi} are constants at the initial reservoir conditions. $W_{\!q}$ is the net water influx and therefore contains the water produced and/or injected.

In the absence of a water drive ($\hat{W}_{\!q}$ = o), **Equation 14** indicates a simple linear relationship between the adjusted reservoir pressure p/Z and the cumulative gas production G_p. Extrapolation of this simple curve to $p = O$ yields directly the initial gas in place G, whereas extrapolation to the appropiate p/Z ratio corresponding to the abandonment reservoir pressure allows determination of the gas reserves.

In the event that water influx is important, **Equation 14** suggests a complex relationship among the variables involved. It likewise suggests a theoretical approach to determine the existence of water influx in the reservoir. If the plot of p/Z versus G_p should indicate a curling tendency as shown in Figure 2.3, this would imply the increasing importance of the $\hat{W}_{\!q}$ term in **Equation 14** which, of course, reflects water encroachment. However, because $\hat{W}_{\!q}$ represents the hydrocarbon pore volume of net encroached water, its value will normally be very small compared to unity, during the early life of the reservoir even though water influx may be effective. This fact, together with the random errors inherent in field data may mask early detection of a water drive in gas reservoirs.

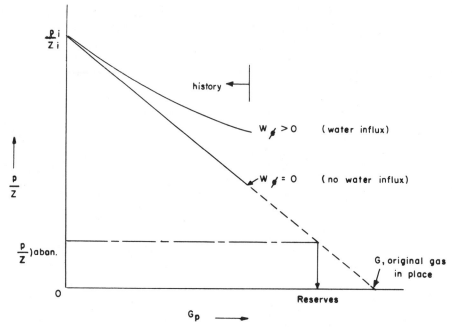

Figure 2.3. Schematic of reservoir pressure vs. cumulative gas produced for a dry gas reservoir.

Depletion Drive Reservoirs

This model constitutes the simplest type of oil reservoir. Essentially, there is no gas cap ($m = 0$) and no water drive (W_e, W_p, $= 0$). Since the oil is saturated at reservoir conditions and as pressure declines a free gas saturation develops, it is normal to consider negligible the formation and connate water compressibilities.

With these assumptions, **Equation 8** becomes

$$N = \frac{N_p \left\{ B_t + (R_p - R_{si}) B_g \right\}}{B_t - B_{ti}} \qquad \textbf{16}$$

Note that an undersaturated reservoir with no water drive or pressure maintenance operations becomes a depletion type when the pressure goes below the bubble point.

2.3 Solving the Integral MBE

We will now discuss methods for solving the general MBE, **Equation 8**, section 8.8. In general, two phases of the calculation procedure must be recognized: (1) the period of matching past history, and (2) the period of prediction. For both phases, special forms of the MBE have been developed with a view to facilitating the mechanical operations involved.

During the matching phase there is no criterion that would guarantee the acceptability of the solution, above all since the data to be introduced in the equation is subject to both error of measurement and random errors. In general, the necessary conditions for acceptability are consistency in the value of the unknowns sought over several time increments and agreement with the volumetric calculation and/or decline curve analysis.

The volumetric calculations in themselves are based on geological and petrophysical data of unknown accuracy. Moreover, there may be a substantial quantitative difference between active oil, as reflected in the material balance calculation, and static volumetric oil. On the other hand, the empirical nature of decline curve analysis limits its use as an independent criterion.

With these restrictions in mind (which should not discourage us!), it would seem definitely advantageous to utilize some mathematical artifice that might aid in "forcing" a solution of the MBE. For instance, the requisite of a straight line relationship is a simple and ubiquitous condition.

As suggested by Havleno and Odeh,[3] the general MBE, **Equation 8**, section 2.2, can be written in the following form

$$X_0 = N X_1 + m N \frac{B_{ti}}{B_{gi}} X_2 + (1 + m) N \frac{B_{ti}}{1 - S_{wi}} X_3$$

$$+ B_1 \Sigma_n W_D (t_D - t_{Dn}) \delta p_n \qquad\qquad 1$$

where the X_I's are all pressure functions, defined as follows

$$X_0 = N_p \{B_t + (R_p - R_{si}) B_g\} + B_w W_p \qquad\qquad 2$$

$$X_1 = B_t - B_{ti} \qquad\qquad 3$$

$$X_2 = B_g - B_{gi} \qquad\qquad 4$$

$$X_3 = (c_f + S_{wi} c_w) \Delta p \qquad\qquad 5$$

and the water influx term has been approximated by the transient form, namely

$$W_e = B_1 \Sigma_n W_D (t_D - t_{Dn}) \delta p_n \qquad\qquad 6$$

where B_1 is the so-called water influx constant, W_D and t_D are the dimensionless water influx function and dimensionless time variables, respectively.

For an oil reservoir with initial gas cap and without water drive (the formation and connate water compressibilities are usually insignificant in saturated reservoirs), **Equation 1** becomes

$$X_0 = N(X_1 + HX_2) \qquad\qquad 7$$

where

$$H = m\,\frac{B_{ti}}{B_{gi}} \qquad\qquad 8$$

If the gas cap volume is known, say, from geologic data, then a graph of X_0 versus $(X_1 + HX_2)$ is a straight line through the origin and with slope N, the initial oil in place. On the other hand if the gas cap volume is not known (that is, both N and m are unknowns), one can plot X_0/X_1 versus X_2/X_1, which yields a straight line with intercept N and slope H wherein m is defined. Or we could have used the previous graph of X_0 versus $(X_1 + HX_2)$, varying H until a straight line through the origin obtains. The extension of the method to other combinations of **Equation 1** is obvious.

It should be pointed out, however, that for water drive reservoirs the attainment of a straight line relationship depends critically on the water influx function assumed, and therefore requires some exploratory work.

Here again, because a great deal of uncertain information goes into the MBE, deterministic methods of analysis are not the most appropriate.

Although several techniques for making uncertainty analyses are available, history-matching with the integral form of the MBE is particularly suited to the methods of least squares or linear programming since the data representing past history gives rise to an overdetermined system of linear algebraic equations. These methods essentially seek a solution which has a minimum error. The linear programming approach theoretically offers a distinct advantage over that of least squares since the residuals in the MBE's tend to be large. However its application requires the use of a computer.

Briefly, in the language of reservoir engineering, the solution of the linear programming problem is defined to be that choice of the unknowns (e.g. N, m) which makes the residuals of the system of linear equations generated by **Equation 8**, section 2.2 minimal. An additional constraint is that N and m must be non-negative, which is always true. For details of the technique, the reader should refer to any of several texts available.

In as much as the least squares method has the advantage of simplicity and yields unbiased estimates of the values sought, it is widely practiced. For a linear empirical model of the form

$$y = a_1 + a_2 x \qquad\qquad 9$$

the criterion of least squares gives for the constants a_1 and a_2

$$a_1 = \frac{\Sigma x_i^2 \, \Sigma y_i - \Sigma x_i \, \Sigma (x_i y_i)}{n \, \Sigma x_i^2 - (\Sigma x_i)^2} \qquad\qquad 10$$

$$a_2 = \frac{n \, \Sigma (x_i y_i) - \Sigma x_i \, \Sigma y_i}{\text{idem}}$$

where n is the number of observations.

One of the fundamental assumptions underlying the estimation procedure that leads to **Equation 10** is that only the dependent variable y is random. That is, the values of x are not random variables. Unfortunately, although MBE models are usually of the form of **Equation 9**, both the dependent and independent variables are random (see Example 1). The estimation of coefficients under these circumstances is quite difficult even for the single linear model of **Equation 9**. In fact, generally applicable techniques for this problem have yet to be devised. It is important that the engineer be aware of this detail and not place blind faith in the results of **Equation 10**.

During the prediction phase of the calculation, two arrangements of the general MBE are generally used: (1) the incremental method, and (2) the Tracy method.[11] In the former, **Equation 8**, section 2.2, is put in incremental or finite difference form and is set up for the beginning and end of a small increment of time (or production) and the difference of the resulting equations is taken, giving, for instance, δN_p or δG_p as a function of the other quantities at the times on pressure bounding the small interval.

For an increment of oil production, δN_p, from a saturated reservoir, **Equation 8**, section 2.2, becomes

$$\delta N_p = \qquad\qquad 11$$

$$\frac{N\delta \left\{ B_t - B_{ti} + m\, B_{ti} \left(\dfrac{B_g - B_{gi}}{B_{gi}} \right) \right\} - N_p\delta \left\{ B_t + (R_p - R_{si})\, B_g \right\} + \delta\,(W_e - W_p B_w)}{B_t + (R_p - R_{si})\, B_g}$$

wherein the second order differentials, $\delta N_p \delta$, have been neglected.

Dividing **Equation 11** by δN_p and separating the various terms of the numerator, one can define different drive indices as suggested by Pirson[8], namely

$$\text{Depletion drive index} = \frac{(N - N_p)\,\delta\left(\dfrac{B_o}{B_g} - R_s\right) - N\,B_{oi}\,\delta\left(\dfrac{1}{B_g}\right)}{\left\{\dfrac{B_o}{B_g} - R_s + R_{av}\right\}\delta N_p}$$

$$\text{Segregation drive index} = -\frac{N_m\,B_{oi}\,\delta\left(\dfrac{1}{B_g}\right)}{\text{idem}}$$

$$\text{Water drive index} = \frac{\delta\,(W_e - W_p\,B_w)/B_g}{\text{idem}} \qquad\qquad 12$$

where R_{av} is the average producing GOR in the interval. Pirson's idea was to predict reservoir performance by simply extrapolating these indices. Obviously, the sum of the indices equals unity.

Another interesting approach to solving the MBE was proposed by Tracy.[11] Instead of estimating the incremental oil production, Tracy suggests estimating the producing gas-oil ratio, R, which is less sensitive to small inaccuracies. Again, the only self-checking feature of the method is that of the original oil in place.

To outline the Tracy method, let's consider a saturated oil reservoir with gas cap and water drive. The controlling MBE can now be written as

$$N = N_p\,\Phi_n + G_p\,\Phi_g - W_\ell\,\Phi_w \qquad\qquad 13$$

where the Φ's are pressure functions defined as

$$\Phi_n = \frac{B_o - R_s\,B_g}{B_t - B_{ti} + m\,B_{ti}\,(B_g - B_{gi})/B_{gi}}$$

$$\Phi_g = \frac{B_g}{\text{idem}}$$

$$\Phi_w = \frac{\Phi_g}{B_g} = \frac{1}{\text{idem}}\,,\ B_w = 1$$

$\left.\begin{array}{c} \\ \\ \\ \\ \\ \\ \\ \end{array}\right\}\quad 14$

The general behavior of the Φ's are shown in Figure 2.4. These functions are only calculated once for the particular reservoir and the prediction computations are easily carried out, usually for unit oil in place.

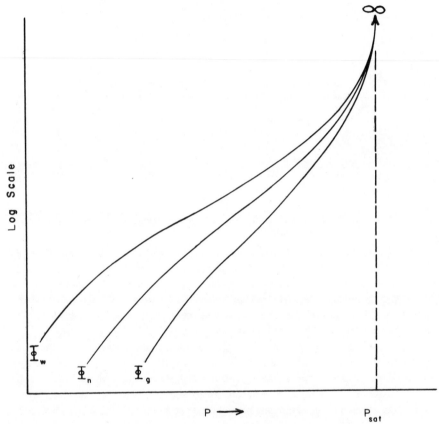

Figure 2.4. Behavior of the Tracy pressure functions.

Equation 13, in prediction form for the interval $j - k$, is

$$1 = (N_{pj} + \Delta_{jk} q_o \, \Delta t) \, \Phi_{nk} + (G_{pj} + \Delta_{jk} R q_o \, \Delta t) \, \Phi_{gk}$$

$$- \left\{ W_{ej} \begin{array}{c} + \; I\,(p,t)\,\Delta t \\ \text{or} \\ - \; I\,(p,t)\,\Delta p \end{array} \right\} \Phi_{wk} \qquad \qquad \textbf{15}$$

I being an average integrand of the water invasion law over the interval $j - k$.

One should not be left with the impression, however, that the solution of **Equation 13** is a direct arithmetic calculation. This depends upon the prediction criteria that are to be incorporated into the solution. It must be remembered that since the MBE contains the parameter gas-oil ratio, R_p, which is dependent both on pressure and saturation of the fluid phases flowing in the reservoir, auxiliary equations are required to completely formulate the problem.

For the depletion model problem, the complementary equation correlating producing gas-oil ratio and relative permeability (saturation) is

$$R = R_s + \lambda \frac{B_o}{B_g} \qquad\qquad 16$$

and that correlating saturation and the explicit parameters (N, N_p) in the MBE, is

$$S_o = \left\{ 1 - \frac{N_p}{N} \right\} \frac{B_o (1 - S_{wi})}{B_{oi}} \qquad\qquad 17$$

This saturation equation follows from the general saturation **Equation 9**, section 2.4, derived in a later section.

Evidently, for predictions, if the gas-oil ratios, R, are assumed at set values, the solution of **Equation 13** is direct. On the other hand, if the behavior of the reservoir is not controlled as such, the set of **Equations 13, 14, 16** and **17** must be solved simultaneously by trial and error. In this case the incremental form, **Equation 15**, is imperative; usually a pressure drop Δp is assumed and $\delta N_p = q \; \Delta t$ is solved for by trial. The details of the calculation procedure are demonstrated in other texts.[2]

Finally, we must emphasize that the foregoing discussion is applicable only to the depletion model problem. If there is a natural water drive the influx function must be specified, thereby adding an additional equation. By the same token, the relevant saturation equations must be derived. These topics are emphasized in subsequent sections of this chapter.

Example 1

A gas field with an active water drive showed a pressure decline from 3000 to 2000 psia over a ten-month period. From the following production data, match the past history and calculate the original hydrocarbon gas in the reservoir. Assume $Z = 0.8$ in the range of reservoir pressures and $T = 600°\,R$.

Data:

t, months	0	2.5	5.0	7.5	10
p, psia	3000	2750	2500	2250	2000
G_p, MMSCF	0	97.6	218.9	355.4	500

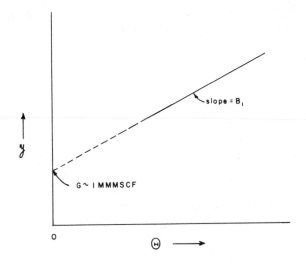

Figure 2.5. Linear relationship between $y = \dfrac{B_g G_p}{B_g - B_{gi}}$ and $\Theta = \dfrac{S_o t_2}{B_g - B_{gi}}$

Solution: Because of the linear pressure decline with time, we will assume that Schilthuis' steady state water influx equation[2] is valid, namely

$$W_e = B_1 \int_o^{tn} \Delta p_n \, dt \qquad\qquad 18$$

where B_1 is a proportionality (aquifer) constant and Δp_n is the cumulative pressure drop from time zero to t_n.

For the above production data, $\Delta p_n = 100t$, which when substituted in **Equation 18**, gives

$$W_e = B_1 \int_o^t 100t \, dt = 50 B_1 t^2 \qquad\qquad 19$$

The MBE, **Equation 12**, section 2.2, is now written as

$$\frac{G_p B_g}{B_g - B_{gi}} = \frac{W_e}{B_g - B_{gi}} + G \qquad\qquad 20$$

which, upon substitution of **Equation 19**, becomes

$$\frac{G_p B_g}{B_g - B_{gi}} = \frac{50 B_1 t^2}{B_g - B_{gi}} + G \qquad\qquad 21$$

This form of the MBE indicates a straight line relationship between $y = \dfrac{G_p B_g}{B_g - B_{gi}}$

and $\Theta = \dfrac{50t^2}{B_g - B_{gi}}$. Such a plot is shown in Figure 2.5, from which the slope is calculated to be B_1 = 235 ft^3/month/psi. Moreover, since the linear relationship is good, we conclude that the steady state water influx equation is a satisfactory representation of the past aquifer behavior.

To conclude the solution of this example, we back extrapolate the curve of Figure 2.5 to t = o, thereby obtaining the original hydrocarbon pore volume, G = 1 MMMSCF.

Example 2

The following information is adapted from a reservoir analysis by van Everdingen et al.[12] The authors have evaluated several values of the water integral, defined here as I assuming that the reservoir-aquifer system's behavior can be represented by the radial-infinite solution. Test the validity of this assumption and determine the original oil in place N and the water influx constant B_1.

Data:

Year	Y	IΦ_w	$\dfrac{1}{\Phi_w}$	ϕ	I	Φ_w
1941				3800		
42				3700		
43	66.0	0.221	0.02432	3500	0.005377	41.12
44	78.8	.289	.04534	3375	.01312	22.05
45	92.3	.367	.06204	3290	.02277	16.12
46	111.0	.461	.07056	3230	.03254	14.17
47	112.3	.472	.0927	3100	.04378	10.79
48	117.0	.498	.1143	3000	.05694	8.79
49	148.2	.670	.1040	3170	.06964	9.62
50	167.5	.782	.1044	3170	.08166	9.58

The values of Y are given in millions and all numbers correspond to the end of the year given. Thus, 3500 psia refers to the end of two years, etc. The bubble point pressure is taken to be 3700 psia.

Solution: Take the MBE in the form

$$N = N_p \Phi_n + G_p \Phi_g - (W_e - W_p) \Phi_w \tag{22}$$

which can be rearranged to

$$Y = N + B_1 I \Phi_w \tag{23}$$

where

$$Y = N_p \Phi_n + G_p \Phi_g + W_p \Phi_w \tag{24}$$

$$W_e = B_1 \Sigma_n W_D \delta \Delta p_n = B_1 I \tag{25}$$

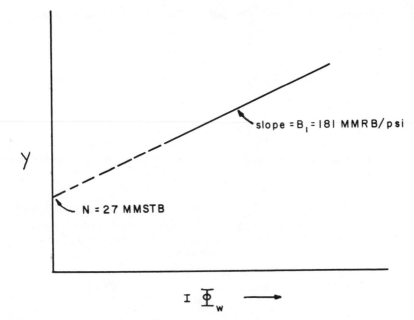

Figure 2.6. A graph of Y vs $I\Phi_w$.

A graph of Y versus $I\ \Phi_w$ should be linear with slope B_1 and intercept N. Figure 2.6 shows such a plot which yields $N = 27\text{MMSTB}$ and $B_1 = 181$ MMRB/psia.

2.4 Gravity Segregation

It has long been recognized[5] that three forms of mechanical energy aid in the expulsion of oil from reservoirs: (1) gas, both dissolved in the oil and free in gas caps, (2) encroaching edge or bottom water, and (3) gravity drainage or self-propulsion of the oil downward in the reservoir rock.

Unfortunately these three forces interfere with each other rather than cooperate and usually only one can be used effectively at one time at the same place in the reservoir. Which one of these agents is locally dominant depends upon local reservoir and operating conditions. Thus, it is possible to have each of the three forces operating individually in different parts of the same reservoir at the same time, or for the relative importance of the forces to vary from time to time.

In distinction from dissolved or dispersed gas-drive conditions where oil and free gas become intermingled, with the oil being propelled in any direction by the gas, the purpose of gravity segregation is to promote a counterflow of the two phases thereby increasing the mass of gas in the gas cap while withdrawing the oil from below the gas-oil contact. In practice, only a relative movement, as compared to counterflow, of the heavier phase downward is accomplished since production rates must remain economical.

Effective application of gravity drive principles means controlling the velocity of gas flow so that it does not entrain the oil but allows it to flow downward and displace the gas which in turn moves up the structure. This critical gas velocity of counterflow essentially controls the rate of oil extraction. Though it will not be uniform throughout the reservoir, in general, good vertical permeability, high dips and low oil viscosity are all favorable for counterflow. Obviously, too, oil can be produced most rapidly from a reservoir in which pressures are maintained by gas injection and still have gravity segregation.

This is so, because as a whole, the counterflow velocity will be controlled by the rate at which pressures are reduced since evolved gas will retard the downward flow of oil. It would appear, then, that the most favorable conditions for fast rates of oil production would be to maintain pressures above solution pressures so that no gas would evolve and there would be no counterflow of gas and oil and the gas movement of the expanding gas cap would be downward with the oil.

If the reservoir has a strong water influx, it would be necessary to maintain a gas cap pressure equal to the aquifer pressure to retard the water encroaching the reservoir, which involves a choice between as to which of the two mechanisms, water drive or gravity, would be more effective.

Because of the complex hydrodynamics of the gravity segregation process, the material balance approach for calculating reservoir performance must require very simplifying assumptions, in particular with regard to the movement of the gas-oil contact and the saturation differences above and below this contact. In addition, if the reservoir pressure is declining during the production phase, the problem acquires another order of complexity which further requires more simplifying assumptions. It must be kept in mind that the material balance approach has the sole purpose of providing the engineer with a first approximation to the problem and therefore too sophisticated adaptations of the method defeats its effectiveness.

In the following chapter the problem of gravity segregation is approached from a more dynamic point of view, making use of two-phase displacement theory. This section, then, will be devoted to a simplified theory which assumes that the gas cap moves uniformly down structure, that there is no water influx, no positional variation of pressure in the gas cap and of the reservoir properties throughout the reservoir. A further assumption includes no loss of gas-cap gas through the oil producing zone since this would imply a dispersed type gas drive.

The method to be outlined is therefore limited to a gross analytical description of the expansion history of the gas cap as a function only of the cumulative recovery. The time element involved in the displacement of the gas-oil interface is taken into account only indirectly. In a sense, the method at best will give a qualitative picture of the relative importance of gravity in a particular reservoir. The problem is depicted in Figure 2.7.

Based on the assumptions described previously the pertinent material balance equation, deduced from **Equation 8**, section 2.2, is

$$N(B_t - B_{ti}) + Nm\,B_{ti}\;\left(\frac{B_g - B_{gi}}{B_{gi}}\right) \;=\; N_p\{B_t + (R_p - R_{si})\,B_g\} \qquad\qquad 1$$

where, theoretically, the ratio of the gas cap volume to the oil volume, m, is a variable that increases as the gas cap expands and the oil zone shrinks. This requirement, however, is obviated by maintaining m a constant at initial reservoir conditions and making the necessary compensations in the saturation equation that must be solved simultaneously with **Equation 1**.

Since overall reservoir performance involves predicting flowing gas-oil ratios which in turn are determined by relative permeabilities, it is of fundamental importance to develop the corresponding saturation equation for the particular reservoir problem. This is the basis of the discussion in the following section.

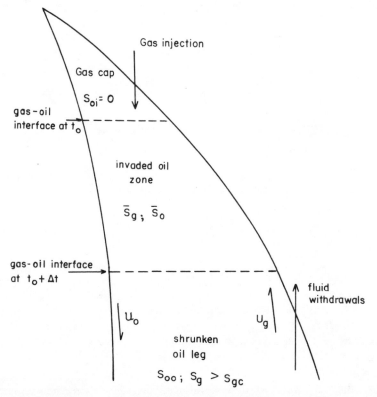

Figure 2.7. Schematic of the gravity segregation mechanism.

Segregation Without Counterflow

If there is a substantial initial gas cap, and/or gas injection, then for average economic production rates, it does not seem likely that counterflow will occur. In deriving the saturation equations under these conditions we may follow Pirson's[8] idea that the gas cap will expand according to Buckley-Leverett displacement theory (see Chapter 3) until the gas saturation in the shrunken oil zone attains its critical level ($Sg \approx 10\%$).

During this period the average oil saturation in the shrunken oil zone, S_{oo}, is given by

$$S_{oo} = \frac{\left\{ \begin{matrix} \text{Total remaining} \\ \text{oil volume} \end{matrix} \right\} - \left\{ \begin{matrix} \text{oil volume in gas cap} \\ \text{expanded zone} \end{matrix} \right\}}{\left\{ \begin{matrix} \text{original oil} \\ \text{pore volume} \end{matrix} \right\} - \left\{ \begin{matrix} \text{pore volume into} \\ \text{which the gas cap has expanded} \end{matrix} \right\}} \qquad 2$$

or

$$S_{oo} = \frac{B_o (N - N_p) - \bar{S}_o \, \Delta V_{cap}/\bar{S}_g}{N B_{ti}/(1 - S_{wi}) - \Delta V_{cap}/\bar{S}_g} \qquad 3$$

where the reservoir gas volume in the zone invaded by the expanding gas cap, ΔV_{cap}, is given by

$$\Delta V_{cap} = \frac{(\text{total expanded gas cap volume}) \, (\text{pore volume})}{(\text{free gas in invaded zone} - \text{liberated associated gas})} \qquad 4$$

or

$$\Delta V_{cap} = \frac{N m B_{ti} \left(\dfrac{B_g - B_{gi}}{B_{gi}} \right)}{1 - \dfrac{\bar{S}_o}{\bar{S}_g} (R_{si} - R_s) \dfrac{B_g}{B_o}} \qquad 5$$

S_g and $\bar{S}_o$ are the average gas and oil saturations, respectively, in the pore volume invaded by the expanding gas cap. These saturation values are obtained by the well known Welge method (see section 3.1) of drawing a tangent from the origin of the fractional gas flow versus saturation curve (f_g versus S_g).

After the gas saturation in the shrunken oil zone has reached its critical value, S_{gc}, we further assume that gas cap expansion ceases and the remaining oil is essentially produced by solution gas drive plus energy supplied by dispersed gas injection, if there exists such a program. Although this latter assumption may seem restrictive, in practice if gravity segregation is effective, producing gas-oil ratios remain fairly constant and low, indicating a small gas saturation in the shrunken oil zone during a large period of the reservoir life.

During the second stage of reservoir performance, in which $S_g > S_{gc}$ in the shrunken oil leg, the previously derived saturation **Equation 3** remains valid. Again the average gas saturation, $\bar{S}_g$, in the invaded zone is that determined by a tangent from the origin in the Welge method, and remains constant. However, the average oil saturation, $\bar{S}_o$ in **Equation 3**, is given by the intercept on the $fg = 1$ line of a tangent with

$$\text{Slope} = \frac{B_{gc} - B_{gi}}{\bar{S}_{gc}(B_g - B_{gi})} \qquad\qquad 6$$

If a standard cumulative volume of gas G_i has been injected, then **Equation 6** becomes

$$\text{Slope} = \frac{\dfrac{B_{gc} - B_{gi}}{B_{gi}} + \dfrac{G_{ic}\,B_{gc}}{Nm\,B_{oi}}}{\bar{S}_{gc}\left\{\dfrac{B_g - B_{gi}}{B_{gi}} + \dfrac{G_i\,B_g}{Nm\,B_{oi}}\right\}} \qquad\qquad 7$$

where the subscript c refers to conditions at the critical gas saturation. Likewise, **Equation 5** remains valid but all variable terms correspond to the pressure at which $S_g = S_{gc}$ in the shrunken oil zone and therefore the value of ΔV_{cap} also remains constant. For the sake of clarity, S_{gc} refers to the critical gas saturation in the shrunken oil zone while $\bar{S}_{gc}$ refers to the average gas saturation in the invaded zone when S_{gc} is attained in the oil zone.

The derivation of **Equation 6** is accomplished by recalling that the tangent in the Welge method is defined as

$$\text{Slope} = \frac{\begin{array}{c}\text{Pore volume through which a}\\ \text{saturation } \bar{S}_{gc} \text{ has travelled}\\ \hline \text{volume throughput}\end{array}}{} = \frac{Nm\,B_{ti}\left.\left(\dfrac{B_{gc} - B_{gi}}{B_{gi}}\right)\right/ \bar{S}_{gc}}{Nm\,B_{ti}\left(\dfrac{B_g - B_{gi}}{B_{gi}}\right)} \qquad 8$$

which simply reduces to **Equation 6**.

In order to generalize the previously derived saturation **Equation 3**, we will finally consider the existence of an active aquifer in the reservoir system. In this case **Equation 3** can be directly extended to give

$$S_{oo} = \frac{B_o\,(N - N_p) - \bar{S}_o\,\Delta V_{cap}/\bar{S}_g - \dfrac{W_{\ell}\,S_{ow}}{S_w - S_{wi}}}{\dfrac{N B_{oi}}{1 - S_{wi}} - \dfrac{\Delta V_{cap}}{\bar{S}_g} - \dfrac{W_{\ell}}{\bar{S}_w - S_{wi}}} \qquad 9$$

where $\bar{S}_{ow}$ and $\bar{S}_w$ are average oil and water saturations, respectively, behind the water front, as obtained by Welge's method.

As stated previously, the object of the saturation equations is to determine relative permeabilities which in turn allow us to predict gas-oil ratios. The gas oil ratio equation, **Equation 17**, section 2.2, is satisfactory for most purposes. However, in high relief reservoirs it may be sometimes warranted to consider gravitational effects. In this event the corresponding producing gas oil ratio is given by

$$R = \frac{\lambda\,B_o}{B_g} + R_s + \frac{\Delta\rho_{o\text{-}g}\,\lambda_g\,Ag\,\sin\alpha}{q_o\,B_g} \qquad 10$$

where α is the formation dip and A, the bulk horizontal area under the expanded gas cap. Theoretically, the relative areas exposed to the well bores must be considered. This, however, would be hard to determine since it is governed by well penetration, perforations, amount of coning and fingering. **Equation 10** is directly obtained from the Darcy flow relationships, **Equations 11** and **12**, given subsequently.

The gravity drainage problem without counterflow is now completely formulated and the calculating procedure may best be demonstrated with an exercise. (See Example 3).

Modifications for Counterflow

Contrary to the previous discussion wherein counterflow was considered negligible, reservoirs with no initial gas cap and no gas reinjection require the occurrence of counterflow if gravity segregation is to be effective. Evidently since counterflow implies the upward movement of evolved gas, then the gas saturation in the oil zone must be well above the critical value.

The Darcy equations for flow of oil and gas are

$$\vec{u}_o = -\lambda_o\,\nabla\cdot(p_o + \rho_o\,gz) \qquad 11$$

$$\vec{u}_g = -\lambda_g\,\nabla\cdot(p_g + \rho_g\,gz) \qquad 12$$

with z positive in the upward direction. We now assume that capillarity is negligible ($p_o = p_g$) and define the net total downward motion of the fluids as

$u_t = u_o + u_g$ where the upwards gas velocity u_g is considered negative. **Equations 11** and **12** for vertical flow can be combined to give

$$u_g = \left\{ u_t + \lambda_o g \Delta \rho_{o \cdot g} \right\} \left\{ 1 + \frac{1}{\lambda} \right\}^{-1} \qquad \text{13}$$

$$u_o = \left\{ u_t - \lambda_g g \Delta \rho_{o \cdot g} \right\} \left\{ 1 + \lambda \right\}^{-1} \qquad \text{14}$$

It is seen from these equations that upward movement of gas occurs if

$$u_t < \lambda_o g \Delta \rho_{o \cdot g} \qquad \text{15}$$

For complete counterflow $u_o = u_g$, $u_t = o$ and **Equations 13** or **14** reduces to

$$u_o = -u_g = \frac{g \Delta \rho_{o \cdot g}}{\left(\dfrac{1}{\lambda} + \dfrac{1}{\lambda_g} \right)} \qquad \text{16}$$

Then, the incremental volume of gas segregated (or oil produced), ΔV_g, is

$$\Delta V_g = -u_g A \Delta t \qquad \text{17}$$

where A is the bulk horizontal area under the previously expanded gas cap. The saturation equation for the shrunken oil zone, by analogy to **Equation 9**, can be shown to be

$$S_{oo} = \frac{B_o (N - N_p) - \left\{ V_g + Nm B_{oi} \left(\dfrac{B_g - B_{gi}}{B_{gi}} \right) \right\} \dfrac{\bar{S}_o}{\bar{S}_g}}{\dfrac{N B_{oi}}{1 - S_{wi}} - \left\{ V_g + Nm B_{oi} \left(\dfrac{B_g - B_{gi}}{B_{gi}} \right) \right\} \dfrac{1}{\bar{S}_g}} \qquad \text{18}$$

Note that the denominator of **Equation 18** is the pore volume of the shrunken oil zone, and $V_g = \Sigma \Delta V_g$.

The calculation procedure may take various approaches. One such approach is outlined as follows:

1. Calculate the reservoir performance in the straight forward manner until the critical gas saturation S_{gc} is attained.
2. Guess ΔN_p and hence Δt since $\Delta N_p = q_o \Delta t$.
3. Find ΔV_g and add to the running summation.
4. Get $\bar{S}_g$ by Welge's method.
5. Calculate S_{oo} from **Equation 18**.
6. Determine the producing GOR using **Equation 16**, section 2.3.
7. Check the simple MBE, **Equation 13**, section 2.3.

Example 3

An oil reservoir with an ititial gas cap, no water drive and no gas reinjection planned, is to be depleted under gravity segregation. Calculate its performance in the absence of counterflow from the original reservoir pressure of 2500 psia. The following data is adapted from Craft and Hawkins[2], p. 382.

Data:

p	B_o	R_s	μ_o/μ_g	$B_g \times 10^4$	Φ_n	$\Phi_g \times 10^2$	k_g/k_o	S_o
2500	1.325	650	56.60	7.96	½	½	–	0.78
2300	1.311	618	61.46	8.43	21.7	2.32	0.004	0.70
1700	1.266	520	81.96	11.36	2.60	0.437	0.010	0.67
1300	1.233	450	102.61	16.16	0.79	0.252	0.028	0.64

$m = 0.3$; $S_{wi} = 0.22$; $S_{gc} = 0.08$; $\bar{S}_g = 0.30$ obtained from a plot of f_g (with segregation terms) versus S_g and refers to the extrapolated value of the tangent from $f_g = 0$.

Solution: The following set of equations is to be solved simultaneously:

$$N = N_p \Phi_n + G_p \Phi_g \qquad\qquad 19$$

$$S_{oo} = \frac{B_o(N - N_p) - \bar{S}_o \Delta V_{cap}/\bar{S}_g}{\{N B_{ti}/(1 - S_{wi})\} - \Delta V_{cap}/\bar{S}_g} \qquad\qquad 20$$

$$\Delta V_{cap} = \frac{Nm B_{ti} \left(\dfrac{B_g - B_{gi}}{B_{gi}}\right)}{1 - \dfrac{\bar{S}_o}{\bar{S}_g} (R_{si} - R_s) \dfrac{B_g}{B_o}} \qquad\qquad 21$$

$$R = \lambda \frac{B_o}{B_g} + R_s \qquad\qquad 22$$

Part 1: Assume $p = 1700$ psia; $R_p = 585$ SCF/STB; $S_{oo} = 0.70$; $N = 1.0$ STB. Then

$$S_o = 1 - S_{wi} - \bar{S}_g = 0.48$$

and

$$\Delta V_{cap} = \frac{(0.30)(1.325) \left(\dfrac{11.36 - 7.96}{7.96}\right)}{1 - \dfrac{0.48}{0.30} \left(\dfrac{130 \times 0.001136}{1.266}\right)} = 0.21$$

also

$$S_{oo} = \frac{1.266\,(1 - N_p) - \dfrac{0.48}{0.30}\,(0.21)}{\dfrac{1.325}{0.78} - \dfrac{0.21}{0.30}} = 0.70\,(\text{inferred})$$

from which $N = 0.1815$

Substituting in the MBE, **Equation 19**, we get

$$1 \overset{?}{=} 0.1815(2.60) + 0.1815(585)\,0.00437 = 0.935$$

At this point, since the MBE is not satisfied within an acceptable tolerance, another pressure value is assumed and the above calculation procedure is repeated. Once the MBE checks satisfactorily, then the cumulative gas-oil ratio must be verified for the assumed value of 585 SCF/STB. The "correct" reservoir pressure corresponding to

$S_{oo} = 0.70$ and $R_p = 585$ is $p = 1753$ psia.

Part 2: In order to continue the performance predictions we now assume: $p = 1300$ psia; $R = 2700$ SCF/STB; $S_{oo} = 0.64$. Recall, from Part 1, that the gas saturation in the shrunken oil zone is now at its critical value and therefore the pressure dependent parameters to be substituted into **Equation 21** for ΔV_{cap} must refer to $p = 1753$ psia. For simplicity, however, the remainder of this example is based on the results obtained at 1700 psia, thereby avoiding interpolation of the basic data.

Therefore $\Delta V_{cap} = 0.21$ as before and remains constant.

Likewise, on the f_g versus S_g curve, we must extrapolate the tangent with

$$\text{Slope} = \frac{B_{gc} - B_{gi}}{S_{gc}\,(B_g - B_{gi})} = \frac{11.36 - 7.96}{0.30\,(16.16 - 7.96)} = 1.382$$

and obtain $\bar{S}_g = 0.40$ at $f_g = 1.0$. Thus

$$\bar{S}_o = 1 - 0.22 - 0.40 = 0.38$$

and

$$S_{oo} = \frac{1.233\,(1 - N_p) - \dfrac{0.38}{0.30}\,(0.21)}{\dfrac{1.325}{0.78} - \dfrac{0.21}{0.30}} = 0.64\,(\text{inferred})$$

from which $N_p = 0.265$

We now substitute these values in the MBE

$$1 \overset{?}{=} 0.265 \, (0.79)$$
$$+ \left\{ (0.1815 \, (585) + (0.265 - 0.1815) \, \left(\frac{585 + 2700}{2} \right) \right\} \, 0.00252$$
$$= 0.823$$

which is highly unsatisfactory. In order to ascertain which of the three initially assumed parameters is more likely in error, we check the gas-oil ratio, namely

$$R = 0.028 \, (102.61) \, \frac{1.233}{0.001616} + 450 = 2650 \text{ SCF/STB}$$

Since this value appears fairly consistent with the inferred value, it would appear that the assumed reservoir pressure should be varied and the calculations of Part 2 repeated.

2.5 Differential Material Balance Equations

Actually there are two ways of formulating the MBE: integral and differential methods. In previous sections the integral form of the MBE was developed whereas this section is devoted to the differential form. The type of MBE chosen for predicting the future performance of a reservoir depends on the amount of past history available and whether or not the gas-oil ratios are to be controlled at given levels or are to be governed directly by the relative permeabilities of the phases; in this latter case the gas-oil ratios are a function of the production which is to be determined.

When the amount of past history is small or the gas-oil ratios are a direct function of the production, the differential form of the MBE offers distinct advantages for solution purposes. In addition, empirical relationships of fluid properties and water influx behavior may be easily incorporated into the MBE and the effects of these parameters on the overall solution can be analyzed rapidly.

In establishing a general differential MBE, essentially the same unified approach, used for the integral form (section 2.2), is followed. In effect we develop the equations corresponding to the three basic conservation variables: expansion, production and influx. The differential MBE, however, is not applicable to undersaturated reservoirs, so that in the derivation that follows the expansion of the connate water and pore volume will be neglected for reasons already explained (section 2.2). Likewise, the treatment of water influx and gas reinjection require special considerations and therefore they will be incorporated after the basic MBE is derived.

$$\text{(System Expansion)} = -\left\{\text{Oil}\right\} + \overbrace{\left\{\begin{array}{c}\text{Associated} \\ \text{Gas}\end{array}\right\} + \left\{\begin{array}{c}\text{Gas Cap} \\ \text{Gas}\end{array}\right\} + \left\{\begin{array}{c}\text{Free} \\ \text{Gas}\end{array}\right\}}^{\text{Expansion of:}}$$

$$= V_p \left\{ -\frac{S_o}{B_o}\frac{d\,B_o}{dp} + \frac{S_o\,B_g}{B_o}\frac{dR_s}{dp} + \frac{m\,(1 - S_{wi})}{B_g}\frac{dB_g}{dp} \right.$$

$$\left. -\frac{S_g}{B_g}\frac{d\,B_g}{dp} \right\} \qquad\qquad 1$$

where V_p is the original reservoir (gas caps plus oil leg) pore volume.

At this point, however, since the MBE is to be expressed in terms of reservoir saturations rather than actual production values, a slight modification of the production and influx terms in the basic conservation **Equation 1**, section 2.2 must be introduced. We now conveniently consider only the oil phase in the reservoir.

$$\{\text{Oil production}\} = \{\text{System expansion}\} \times \left\{\begin{array}{l}\text{Fraction of oil in the} \\ \text{total hydrocarbon flow}\end{array}\right\}$$

$$\{\text{Oil production}\} = \textbf{Equation 1} \times f_o$$

where, in the absence of gas reinjection, gravitational and capillary effects

$$f_o = \{1 + \lambda\}^{-1} \qquad\qquad 2$$

Then,

$$\left\{\begin{array}{l}\text{Rate of oil} \\ \text{desaturation}\end{array}\right\} = \{\text{Oil production}\} + \{\text{Oil shrinkage}\}$$

$$V_p\frac{d\,S_o}{dp} = \textbf{Equation 1}\,f_o + V_p\frac{S_o}{B_o}\frac{d\,B_o}{dp} \qquad\qquad 3$$

which, upon making the appropriate substitution, rearranges to

$$\frac{d\,S_o}{d\,p} = \left\{ S_o \left(\frac{B_g\,d\,R_s}{B_o dp} + \frac{1}{B_g}\frac{d\,B_g}{dp} + \frac{\lambda}{B_o}\frac{d\,B_o}{dp} \right) \right.$$

$$\left. - (1 + m)\,(1 - S_{wi})\,\frac{1}{B_g}\frac{d\,B_g}{dp} \right\} \{1 + \lambda\}^{-1} \qquad\qquad 4$$

In the absence of a gas cap (m = o) **Equation 4** reduces to the zero-dimensional differential **Equation 12**, section 2.7 obtained directly from the beta model. The proof that the material balance approach is a semi-steady state approximation of the dynamic reservoir model of the depletion problem is now complete.

As mentioned previously if gas-oil ratios are controlled at an arbitrary value R^0, then B_g ($R^0 - R_s$)/B_o is substituted for λ in **Equation 4**; but then it is evident that this differential form of the MBE is not the most convenient, unless its use merely involves a slight change in an existing program. On the other hand, if a fraction r of the produced gas is reinjected, the above mobility ratio term must be replaced as follows

$$\lambda \rightarrow (1 - r)\,\lambda - rR_s\,\frac{B_g}{B_o} \qquad\qquad 5$$

When water influx is important, the apparent difficulty in incorporating its effect lies in the fact that the MBE is expressed in terms of saturation values and the aquifer is not directly included in the overall system. This is obviated by including the water encroachment term directly into the system expansion **Equation 1** as $\dfrac{d\widetilde{W}_{\ell}}{dp}$ where $\widetilde{W}_{\ell}$ is the net water influx per unit original hydrocarbon system pore volume. **Equation 4** can now be written in a generalized form as

$$\frac{d\,S_o}{dp} =$$

$$\left\{ S_o \left(\frac{B_g}{B_o}\frac{d\,R_s}{dp} + \frac{1}{B_g}\frac{d\,B_g}{dp} + \frac{\lambda}{B_o}\frac{d\,B_o}{dp} \right) - (1 + m)\,(1 - S_{wi})\,\frac{1}{B_g}\frac{d\,B_g}{dp} \right.$$

$$\left. + \frac{\widetilde{W}_{\ell}}{B_g}\frac{d\,B_g}{dp} - \frac{d\,\widetilde{W}_{\ell}}{dp} \right\} \{1 + \lambda\}^{-1} \qquad\qquad 6$$

where λ may be defined by **Equation 5** if a dispersed type gas reinjection program is to be considered.

It is well to reemphasize some other basic assumptions implicit in **Equation 6**. No oil saturation is assumed present in the gas cap, so that the initial water saturation S_{wi} is the same in the gas as in the oil zone; also any gas reinjection is of the dispersed type. More importantly, the oil saturation S_o refers to the average over the original reservoir (gas cap and oil zone) pore volume.

Therefore, in order to get a realistic picture of the producing gas-oil ratios, we must express the saturation in the oil leg in terms of the average for the reservoir. With the previous assumptions and no allowance for release of gas from the trapped oil due to pressure decline, the relevant saturation equation is the generalized saturation **Equation 9**, section 2.4 previously derived but with all terms divided by the pore volume, $NB_{oi}/(1 - S_{wi})$. Thus we have

$$S_{oo} = \frac{S_o - \dfrac{\bar{S}_o}{\bar{S}_g} \Delta \widetilde{V}_{cap} - \dfrac{\bar{S}_{ow} \widetilde{W}_\ell}{\bar{S}_w - S_{wi}}}{1 - \dfrac{\Delta \widetilde{V}_{cap}}{\bar{S}_g} - \dfrac{\widetilde{W}_\ell}{\bar{S}_w - S_{wi}}}$$

7

where the terms with the tilde refer to unit original pore volume. $\Delta \widetilde{V}_{cap}$ is defined as before by **Equation 5**, section 2.4, but corrected to a pore volume basis.

A somewhat more versatile differential MBE can be obtained by directly differentiating the Tracy arrangement of the integral MBE, namely

$$N = N_p \, \Phi_n + G_p \, \Phi_g - W_\ell \, \Phi_w$$

8

Because N is a constant, differentiation gives

$$- \left\{ N_p \, \frac{d \, \Phi_n}{dp} + (G_p) \, \frac{d \, \Phi_g}{dp} - W_\ell \, \frac{d \, \Phi_w}{dp} \right\} \, dp$$

$$= \left\{ \Phi_n \, q_o + \Phi_g \, Rq_o + \Phi_w \, \frac{d \, W_\ell}{dt} \right\} \, dt$$

9

where

$$q_o = \frac{d \, N_p}{dt}$$

10

$$Rq_o = \frac{d \, G_p}{dt}$$

11

and

$$G_p = \frac{(1 - N_p \, \Phi_n)}{\Phi_g} \, , N = 1$$

12

This form of the MBE has the advantage that a variable oil rate q_o (using increments of N_p instead of decrements of pressure) is easily incorporated into the calculations, giving performance predictions that might be used in optimizing a production schedule with water influx and gas reinjection.[10].

For the differential Tracy MBE, the corresponding saturation equation is best expressed in terms of production. Thus, the generalized saturation **Equation 9**, section 2.4, is again valid, with $N = 1$.

As mentioned previously, the saturation **Equation 9**, section 2.4, does not allow for release of gas from the trapped oil at low pressures. Although this may not occur to any great extent, the MBE assumes that it does. The effect is an increased water invasion into the oil zone, so that a consistent saturation equation should be adjusted for this. To do this, it is necessary to keep a running account of dissolved gas trapped at various pressures. If N_{tr} represents trapped oil and $W_e{}^*$ pore volume encroached, then for a pressure decrement Δp,

$$\Delta \widetilde{W}_{\not{e}}{}^* = \frac{\Delta W_e + N_{tr}\Delta p B_g \dfrac{dR_s}{dp}}{\overline{S}_w - S_{wi}}$$

13

and

$$\Delta N_{tr} = \Delta \widetilde{W}_e{}^* (1 - \overline{S}_w)/B_o$$

14

$\widetilde{W}_e{}^*$ is calculated from the running summation of $\Delta W_{\not{e}}$ and ΔN_{tr} (up to the current Δp);   and B_g need only be approximate, and a summation of $\Delta \widetilde{W}_e{}^*$ is kept. **Equation 9**, section 2.4, can now be written as

$$S_{oo} = \frac{B_o(1 - N_p - N_{tr}) - \dfrac{\overline{S}_o}{\overline{S}_g} \Delta V_{cap}}{\dfrac{B_{oi}}{1 - S_{wi}} - \widetilde{W}_e{}^* - \dfrac{\Delta V_{cap}}{\overline{S}_g}}$$

15

with $W_{\not{e}}$ now based on $N = 1$ or on unit pore volume, the latter as required by the Muskat differential form. $\widetilde{W}_e{}^* = \Sigma \Delta \widetilde{W}_e{}^*$.

2.6 Solving the Differential MBE

Interestingly, both forms of the differential MBE presented in the latter section, **Equations 6** and 9, section 2.5, represent the general differential equation

$$\frac{dy}{dx} = f(x,y)$$

1

where the variables are not separable. This class of equation is particularly suitable for well known numerical techniques such as, among others, those of Runge-Kutta or Hamming. Moreover, both of the above mentioned MBE's have been found to be very stable, thereby allowing large decrements of the independent variable to be used without significant deterioration in results. For hand computations even the simple Euler integration formula may be satisfactorily used with pressure decrements of the order of 200 psi.

The particular form of the MBE, namely **Equation 9**, section 2.5, not only has the advantage of handling directly varying production schedules, but is very suitable for matching past performance and, of course, determination of unknowns such as the original oil in place N and the water influx constant. Since these unknowns are constant coefficients of the differential MBE, we may use the least squares estimation procedure to optimize the values of these parameters. Several such techniques are available for optimization both in the time domain and in the Laplace transform domain. Himmelblau[4] gives an excellent review of these methods.

Moreover, for both matching and prediction purposes, a simple but powerful tool is the digital simulators of analog computers such as MIMIC, CSMP, etc.

Finally, it is apparent that some empirical equations for the various functions and their derivatives would be useful in the differential MBE's. Rather than use general high-order polynomials, as has become popular since the availability of computers, some simple and accurate empirical forms which likewise facilitate taking their derivatives will be suggested. The equations for fluid properties are necessarily limited to phase equilibrium below the bubble point. The pressure dependent parameters are amenable to the following models[10]

$$R_s, B_o \quad = \quad b_o + b_1 p - b_2 \exp(-b_3 p) \tag{2}$$

$$\mu_o/\mu_g \quad = \quad \begin{cases} \dfrac{b_o}{b_1 + p} + b_2, \text{ or} \\ b_o + b_1 \exp(-b_2 p) \end{cases} \tag{3}$$

$$B_g^{-1}\frac{dB_g}{dp} = \begin{cases} 1/p, \text{ or} \\ 1/p + b_o - b_1 p \end{cases} \tag{4}$$

$$B_g/B_o \quad = \quad \begin{cases} b_o/p - b_1, \text{ or} \\ b_o/p - b_1 + b_2 p \end{cases} \tag{5}$$

$$\Phi_g, \Phi_w = \quad b_o p + b_1 p/(p_i - p) \tag{6}$$

$$\Phi_n \quad = \quad (b_o p + b_1 p^2)/(p_i - p) \tag{7}$$

The relative permeability ratio (gas/oil) may conveniently be obtained from Corey's[1] equations for drainage displacement mechanisms, as characterized by the gas drive. These equations are

$$k_{ro} = (1 - S)^4 \tag{8}$$
$$k_{rg} = S^3(2 - S)$$

where

$$S = S_g/(1 - S_{wi}) \tag{9}$$

For the k_g/k_o ratio, some simpler and quite flexible power forms are

$$k_g/k_o = \begin{cases} b_o \left(\dfrac{S_g - b_1}{S_o} \right)^n , \text{ or} \\ b_o \left(\dfrac{S_g}{S_o} \right)^n \end{cases} \qquad n = 3 \text{ or } 4 \qquad \qquad 10$$

In the case of water encroachment the Welge method requires a knowledge of the k_w/k_o ratio. The individual permeability values may be obtained from the Naar-Henderson[7] imbibition equations

$$k_{ro} = (1 - 2S)^{3/2} \left\{ 2 - (1 - 2S)^{1/2} \right\}$$

$$k_{rw} = S^4 \qquad \qquad 11$$

where

$$S = (S_w - S_{wi})/(1 - S_{wi}) \qquad \qquad 12$$

For the k_w/k_o ratio a fairly accurate estimate can be had drawing a straight line on a semi-log graph through the two points

$$\log (k_w/k_o) = \begin{cases} -3 \text{ at } S_{or} \\ -2 \text{ at } S_{wi} \end{cases} \qquad \qquad 13$$

and then using the following relationship

$$k_w/k_o = b_o \exp (b_1 S_w) \qquad \qquad 14$$

Example 4

A saturated oil reservoir with an original pressure of 3500 psia produced 1.9 MMSTB of oil as the reservoir pressure declined to 3000 psia. Calculate the original oil in place using the tabulated PVT data. Predict the reservoir performance assuming gravity segregation with counterflow and without gas injection. Will gravity segregation be effective at an economic off-take rate of 5500 STB/D?

Data:

p, psia	$B_g \times 10^4$, $\dfrac{RB}{STB}$	B_o, $\dfrac{RB}{STB}$	R_s, $\dfrac{SCF}{STB}$	μ_g, cps	μ_o, cps
500	6.70	1.565	1028	0.02390	0.486
3250	7.21	1.532	958	.02295	.500
3000	7.81	1.500	888	.02200	.513
2800	8.38	1.474	832	.02124	.526
2600	9.03	1.448	776	.02048	.537
2500	9.38	1.435	748	.02010	.544

Y_o = 0.58 (H_2O = 1); 33.3° API

Y_g = 0.78 (air = 1)

Dip = 10°

N_p to date 1.9 MMSTB;

Present pressure 3000 psia;

h = 15 ft; ϕ = 0.20

S_{wi} = 0.10; Area = 5000 acres

Width = 10,000 ft; k = 1 darcy

Pore volume, V_p = 116 MMRB

Cumulative GOR to date = 950 SCF/STB

Analysis of behavior to date: From MBE, **Equation 16**, section 2.2:

$$
\begin{aligned}
N &= \frac{N_p(B_o - B_g(R_s - R_p))}{B_g(R_{si} - R_s) - B_{oi} + B_o} \\
&= \frac{1.9(1.500 - 0.000781(888 - 950))}{0.000781(140) - 0.065} \\
&= 66.5 \text{ MMSTB}
\end{aligned}
$$

$$
N \text{ (volumetric)} = V_p\, S_{oi}/B_{oi} = \frac{5000(15)\,(0.2)\,(0.9)\,(7758)}{1.565} = 66.7 \text{ MMSTB}
$$

Although the difference in the calculated N's is small, it is not enough to determine whether there was an initial gas cap or a water drive, because of the sensitivity of the method to small errors in data. By Muskat's MBE, **Equation 4**, section 2.5.

$$
\Delta S_o = f_o \left\{ S_o \left(\frac{B_g dR_s}{B_o dp} + \frac{\lambda dB_o}{B_o dp} \right) - \frac{S_g}{B_g} \frac{dB_g}{dp} \right\}_{ave} \Delta p \tag{15}
$$

$$
f_o = (1 + \lambda)^{-1} \tag{16}
$$

Note that the quantities in brackets are averages over the interval for the Euler integration method, hence we calculate them at p_{av} = 3250 psia. Further, assume k_{rg} = o, therefore f_o = 1. From production data at p = 3000 psia

$$
S_o = \frac{(N - N_p)B_o}{V_p} = \frac{(66.7 - 1.9)1.500}{116} = 0.838
$$

thus, at p = 3250 psia

$$
S_o \sim 1/2(0.90 + 0.838) = 0.869
$$

By Muskat's equation

$$
\Delta S_o = \left\{ 0.869 \left(\frac{0.000721 \times 0.28}{1.532} \right) + \frac{0.031}{3250} \right\} .500 = 0.0619
$$

Therefore at p = 3000 psia, S_o = 0.838, which agrees with both the volumetric and production calculations. This would seem to indicate little or no water drive or initial gas cap as was previously assumed.

Prediction without injection but assuming segregation by counterflow:
Assume 4 MMSTB to be produced in the next two years at the end of which p = 2600 psia, k_{rg} = .00222 and k_{ro} = 0.66 at S_o = 0.80.

At p = 2600 psia, $S_o = \dfrac{(N - N_p)B_o}{V_p} = \dfrac{(66.7 - 5.9)\,1.448}{116} = 0.759$

At the average interval pressure, p = 2800 psia:

$$P_o = \frac{350V_o + 0.0764V_gR_s}{5.615B_o} = \frac{350(0.58) + 00764(0.78)832}{5.615(1.474)} = 42.3\,\frac{\#}{ft^3}$$

$$P_g = \frac{0.0764V_g}{5.615B_g} = \frac{0.0764(0.78)}{5.615(0.000838)} = 12.5\frac{\#}{ft^3}$$

From the counterflow **Equations 13** and **17**, section 2.4, in field units

$$\Delta V_g = 0.007825 k\Delta t \left(\frac{\mu_o}{k_{ro}} + \frac{\mu_g}{k_{rg}}\right)^{-1} \Delta P_{o\text{-}g} A \, \text{Sin } a$$

$$\Delta V_g = 0.007825 \times 1 \times 730 \left(\frac{0.526}{0.66} + \frac{0.212}{0.00222}\right) 29.8 \times 15 \times 10^4 \times 0.1736$$

= 0.441 MMRB of gas segregated.

Now k_g/k_o in MBE is determined by the saturation distribution in the shrunken oil zone, S_{oo}, if wells in the secondary gas cap and/or water-invaded zone, if any, are shut in. This is related to S_o which is based on the original oil zone by the relation, **Equation 7**, section 2.5.

$$S_{oo} = \frac{S_o - \widetilde{V}_g \dfrac{\overline{S}_o}{\overline{S}_g}}{1 - \dfrac{V_g}{\overline{S}_g}} = \frac{0.80 - \dfrac{0.441}{116} - \dfrac{0.47}{0.43}}{1 - \dfrac{0.441}{116 \times 0.43}} = 0.801$$

In this case we let $\widetilde{W}_q$ = o and the average gas saturation in the secondary gas cap, $\overline{S}_g$ = 0.43 from displacement theory. Since the value of S_{oo} = 0.801 is quite close to S_o = .80, this latter value, first selected, will be used to obtain the relative permeability values required (k_g/k_o = .0034) by the MBE.

From previous calculations, $\Delta S_o = 0.838 - 0.759 = 0.079$. Therefore from **Equation 15**

$$\Delta S_o = 0.079 = \left\{ 1 + 0.0034 \left(\frac{0.526}{0.0212}\right) \right\}^{-1} \left\{ 0.80\left[\left(\frac{0.000838 \times 0.28}{1.474}\right)\right.\right.$$
$$\left.\left. + 0.0034 \left(\frac{0.526}{0.02124}\right)\left(\frac{0.00013}{1.474}\right)\right] + \frac{0.1}{2800} \right\} \Delta p$$

Whence $\Delta p = 506$ psia and $p_{2\ yrs} = 2500$ psia as compared with $p_{2\ yrs} = 2600$ psia assumed.

Thus, at $p = 2500$ psia, the corrected $S_o = \dfrac{60.8\ (1.435)}{116} = 0.753$

Gas originally in place $= NR_{si} = 66.7\ (1.028) = 68.5$ MMMSCF
Gas produced to $p = 3000$ psia $= N_p R_p = 1.9\ (950) = 1.8$ MMMSCF
Gas remaining at $p = 3000$ psia $= 66.7$ MMMSCF
Gas remaining at $p = 2500$ psia $= 60.8\ (748) + 116\ (0.147)/0.000938 = 63.7$ MMMSCF

Therefore, average gas-oil ratio, $R = \dfrac{66.7 - 63.7}{4} = 750$ SCF/STB.

Actually the preceding computations should be repeated with a better guess on pressure. The feasibility that a predetermined economical rate of oil production of 5500 STB/D will give significant gravity segregation in this reservoir is analyzed in Example 5 of Chapter 3.

2.N Closure

This chapter describes some of the conventional approaches to reservoir history-matching and performance forecasting, using both the integral and differential forms of the material balance equations. The philosophy has been to indicate the usual approaches for analyzing reservoir behavior with the idea that the reader is generally familiar with the analytical and graphical techniques that are available for these problems.

A brief comparison between the material balance method and dynamic reservoir simulators is presented with a view to demonstrate the important fact that the MBE's are in effect, semi-steady state representations of the true dynamic behavior of the reservoir. Later on, the MBE's in integral and differential form are derived using a unified approach which provides considerable insight into the general problem, so that the engineer can literally build his particular mathematical model for the specific reservoir. A similar approach is used in deriving a general saturation equation which can easily be adapted to the material balance model being used.

Special emphasis is given to the behavior of reservoirs producing under combination drives, in particular, dispersed gas injection and gravity segregation, the latter with and without counterflow. In order to aid in the application of the theory, several field type examples are demonstrated.

Finally, several simple empirical equations are suggested which should help considerably in relating the behavior of various reservoir parameters.

Questions

1. How do capillary and viscous forces affect gravity segregation?

2. Show that $\dfrac{1}{B_g}\dfrac{dB_g}{dp}=\dfrac{1}{p}$ for Z constant.

3. Derive the following gas balance for a reservoir producing under gravity segregation

$$G_p = NR_{si} - (N - N_p) R_s + Nm\frac{B_{oi}}{B_g}$$

$$-\frac{1}{B_g}\left[V_g + \frac{S_{go}B_{oi}N}{1 - S_{wi}} - \frac{S_{go}}{S_g}\left\{V_g + NmB_{oi}\frac{B_g - B_{gi}}{B_{gi}}\right\}\right]$$

where S_{go} refers to the gas saturation in the oil zone, that is, corresponding to S_{oo}.

4. Modify **Equation 18**, section 2.4, to include water encroachment.

5. Show that for a reservoir with an expanding gas cap from which no gas is produced and into which a cumulative standard volume G_i is reinjected, the counterflow equation is given as

$$U_t = NmB_{oi}\frac{\Delta B_g}{B_{gi}A\Delta t} + \frac{\Delta G_i B_g}{\Delta t}$$

6. Why does pressure maintenance often keep down gas-oil ratios although in the instantaneous gas-oil ratio formula, **Equation 16**, section 2.3, the higher gas solubility and lower gas formation factor would indicate a higher GOR?

7. Derive **Equation 5**, section 2.5. What assumptions are involved?

8. Show how a conformance factor can be introduced into the integral and differential forms of the MBE in order to account for injected gas not contacting all of the reservoir oil.

Symbols (in Field Units)

a_i	empirical constants
A	cross-sectional area under the gas cap, ft^2
A_r	cross-sectional area at the perimeter of the reservoir, ft^2

b_i	empirical constants
B_g	gas formation volume factor, RB/SCF
B_o, B_w	oil and water formation volume factor, respectively, RB/STB
B_t, B_{tw}	two phase formation factor of oil and water, respectively, RB/STB
B_1	water influx constant, RB/psia
c_o, c_g, c_w, c_f	compressibilities of oil, gas, water and formation, respectively, psi^{-1}
c_{oe}	effective oil compressibility, psi^{-1}
exp	e, constant
f_g, f_o	fractional gas and oil flow, dimensionless
G	original gas in place, SCF
G_i	cumulative injected gas, SCF
G_{ic}	cumulative injected gas, with S_{gc} is attained in oil zone, SCF
G_p	cumulative produced gas, SCF
g	gravitational constant, 1/144
h	formation thickness, ft.
H	dimensionless variable, **Equation 8**, section 2.3
I	aquifer integral, e.g. **Equation 25**, section 2.3
k	absolute permeability, millidarcys
k_o, k_g, k_w	effective oil, gas and water permeability, respectively, millidarcys
k_{ro}, k_{rg}, k_{rw}	relative oil, gas and water permeability, dimensionless
log	log base 10
m	ratio of gas cap pore volume to oil leg pore volume
n	empirical constant
N	original oil in place, STB
N_p	cumulative oil produced, STB
N_{tr}	cumulative oil trapped in water invaded zone, STB
p	reservoir pressure, psia
P_c	capillary pressure, psia
p_o, p_g	reservoir pressure of the oil and gas, psia
q_o, q_g	volumetric rate of oil and gas, STB/D or SCF/D
R	producing surface gas-oil ratio, SCF/STB
r	length measurement, ft or fraction of produced gas reinjected
R_p	cumulative gas oil ratio, SCF/STB
R_s	solution gas oil ratio, SCF/STB
R_{si}	initial solution gas-oil ratio, SCF/STB
r_b	radius of hydrocarbon reservoir, ft
S	reduced saturation, **Equation 9** or **12**, section 2.6
S_o, S_g	saturation of oil and gas, respectively, over the entire hydrocarbon pore volume, dimensionless
S_g	average gas saturation in gas cap invaded zone, dimensionless
S_{gc}	critical gas saturation in oil zone, dimensionless

$\bar{S}_{gc}$	average gas saturation in gas cap invaded zone when S_{gc} exists in the oil zone, dimensionless
S_{go}	gas saturation in oil zone, dimensionless
$\bar{S}_o$	average oil saturation in gas cap invaded zone, dimensionless
S_{oo}	oil saturation in oil zone, dimensionless
S_{or}	residual oil saturation, dimensionless
$\bar{S}_{ow}$	average oil saturation in water invaded zone, dimensionless
$\bar{S}_w$	average water saturation in water invaded zone, dimensionless
S_{wi}	initial water saturation in hydrocarbon reservoir, dimensionless
T	reservoir temperature, $^\circ R$
t	time, days
t_D	dimensionless time
u_o, u_g	Darcy velocity of oil and gas phases, respectively, ft/day
u_t	total Darcy phase velocity ($= u_o + u_g$), ft/day
V_{cap}	gas volume in gas cap invaded zone, RB
$\tilde{V}_{cap}$	gas volume in gas cap invaded zone, per unit pore volume, dimensionless
V_g	cumulative volume of gas segregated, SCF
W_D	cumulative water influx function, dimensionless
W_e	cumulative water influx, RB
W_ℓ	net cumulative water influx ($= W_e - W_p$), RB
$\overline{W}_e$	net cumulative water influx per unit original gas pore volume, dimensionless
$\tilde{W}_\ell$	net cumulative water influx per unit original pore volume, dimensionless
W_ℓ^*	pore volume encroached by water, RB
x	arbitrary variable
Y, y	arbitrary variables
Z	gas compressibility factor, dimensionless
z	vertical space coordinate, ft
a	formation dip, degrees
δ	differential operator
Δ	increment
η	hydraulic diffusivity constant, ft^2/day ($= 0.00633k/\phi \mu c$)
ϕ	fractional porosity
Φ_n, Φ_g, Φ_w	Tracy pressure functions for oil, gas and water, dimensionless
γ_o, γ_g	specific gravity of oil and gas, dimensionless
λ	mobility ratio, λ_g/λ_o
λ_o, λ_g	mobility or oil and gas, respectively, millidarcy/cp
μ_o, μ_g	viscosity of oil and gas, cp
∇	nabla or Laplacian operator
ρ_o, ρ_g	density of oil and gas, lb/ft^3;
X_i	Havlena and Odeh's pressure functions, RB/STB

References

1 Corey, A. T. "The Interrelation Between Gas and Oil Relative Permeabilities." *Prod. Monthly* (Nov. 1954), 19 (1), 38.

2 Craft, B. C. and Hawkins, M. F. *Applied Petroleum Reservoir Engineering,* Prentice-Hall, Inc. (1959) New York.

3 Havlena, D. and Odeh, A. S. "The Material Balance as a Straight Line." *Trans. AIME* (1963) 228, 896.

4 Himmelblau, D. M. *Process Analysis by Statistical Methods,* John Wiley and Sons, Inc. (1970).

5 Lewis, J.O. "Gravity Drainage in Oil Fields." *Transactions AIME* (1944), 151, 133.

6 Muskat, M. *Physical Principles of Oil Production,* McGraw Hill (1949), New York, 302.

7 Naar, J. and Henderson, J. H. "An Imbition Model – Its Application to Flow Behavoir and the Prediction of Oil Recovery." *Trans. AIME* (1961).

8 Pirson, S. J. *Elements of Oil Reservoir Engineering,* 2nd Ed., McGraw Hill (1958), New York.

9 Roebuck, I. F. et al. "The Compositional Reservoir Simulator Case 1 – The Linear Model." *Trans. AIME* (1969).

10 Sandrea, R. and Nielsen, R. F. "Combination Drive Predictions by the Muskat and Differential Tracy Material Balances Using Various Empirical Relations and Theoretical Saturation Equations." *J. Can. Pet. Tech.* (July-Sept. 1969).

11 Tracy, G. W. "Simplified Form of the Material Balance Equation." *Trans. AIME* (1955), 204, 243.

12 van Everdingen, A. F. Timmerman, E. H. and McMahon, J. J. "Application of the Material Balance Equation to a Partial Water Drive Reservoir." *Trans. AIME* (1953), 198, 52.

3. Immiscible Gas - Oil Displacement

3.0 Introduction

The previous chapter dealt mainly with the gross movement of fluids in the reservoir under thermodynamic equilibrium conditions, although no mass transfer between chemically similar species was considered. Moreover, no attempt was made to assess the efficiency with which the injected fluid, gas or water, would displace and contact the resident oil, namely, the recovery efficiency.

An engineering evaluation of this so called recovery efficiency of a gas injection project requires a knowledge of three factors: (1) the efficiency with which the gas displaces the oil on a pore scale, (2) the horizontal sweep efficiency which refers to the area contacted, on a horizontal plane, by the advancing injected fluid as a function of time, and (3) the vertical sweep efficiency or injected fluid profile as it advances through layers of different physical and fluid properties.

Interestingly, if we neglect all phase behavior effects between the injected gas and the resident oil and gas phases, in other words, consider the displacement to be only immiscible, all methods for determining these three efficiencies namely, displacement, horizontal and vertical, hinge on one of two basic theories: (1) simultaneous flow of the displacing and resident fluids behind the injection front, or (2) separate flow of each phase, the oil ahead of the front and the gas behind. In the case of displacement efficiency, these two theories are known as Buckley-Leverett and Dietz, respectively. On the other hand, all three efficiencies depend critically on a common parameter, the mobility ratio of the displacing to resident fluids.

It is the object of this chapter to present a consolidated viewpoint for evaluating gas injection projects under dynamic conditions. In particular, both gravitational and capillary forces will now be considered, however, under the common assumption that all fluids involved are incompressible and behave immiscibly.

3.1 The Buckley-Leverett Model

This model, although originally formulated thirty years ago[1], is still the subject of current research and controversy which in a sense makes it fresh and challenging. The Buckley-Leverett (B-L) theory is essentially a one dimensional representation of the simultaneous and parallel flow of the displacing and displaced fluids without introducing the concept of fluid interface. Its extension to multidimensional space is only tractable numerically.

The derivation of the model for the gas-oil displacement problem, is based on the following set of equations:

Continuity:
$$\frac{\partial (f_g u_t)}{\partial x} = - \phi \frac{\partial S_g}{\partial t} \qquad 1$$

Darcy (oil):
$$u_o = - \lambda_o \left(\frac{\partial p_o}{\partial x} - \rho_o g \, \mathrm{Sin} \, \alpha \right) \qquad 2$$

Darcy (gas):
$$u_g = - \lambda_g \left(\frac{\partial p_g}{\partial x} - \rho_g g \, \mathrm{Sin} \, \alpha \right) \qquad 3$$

Capillary:
$$P_c = p_g - p_o \qquad 4$$

Saturation:
$$S_o + S_g + S_{wi} = 1 \qquad 5$$

$$f_g = u_g/u_t \qquad 6$$

$$u_t = u_o + u_g = \text{constant} \qquad 7$$

Combining **Equations 2** through **7** yields the well known fractional flow relationship

$$f_g = \frac{1 + \dfrac{\lambda_o A}{q} \left(\left| \dfrac{dP_c}{dS_g} \dfrac{\partial S_g}{\partial x} \right| - \Delta \rho_{o\text{-}g} g \, \mathrm{Sin} \, \alpha \right)}{1 + \lambda_o/\lambda_g} \qquad 8$$

where u_t, the constant injection velocity, has been replaced by the injection volume rate per unit area at the given point, q/A. The angle of formation dip

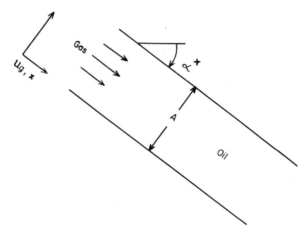

Figure 3.1. Geometry of the Buckley-Leverett Model.

α is positive for upstructure injection (see Figure 3.1) and the absolute sign on the capillary pressure term emphasizes the fact that the capillary forces always increase the fractional flow regardless of the direction of flow of the displacing phase. Note also that P_c is defined only as the pressure difference between phases at some point, without reference to interfacial tension forces. In fact, the P_c-term includes effects other than capillarity.

Since u_t is constant, if we could neglect capillary pressure (as would occur if the reservoir rock was equally wet by oil and gas), then the fractional gas flow would be a function only of the gas saturation, that is, $f_g(S_g)$. Then the continuity **Equation 1** can be written as

$$\frac{q}{\phi A} \frac{df_g}{dS_g} \frac{\partial S_g}{\partial x} + \frac{\partial S_g}{\partial t} = 0 \qquad\qquad 9$$

This is a first order hyperbolic partial differential equation, amenable to the method of characteristics.

The total time derivative of saturation is given as

$$\frac{dS_g}{dt} = \frac{\partial S_g}{\partial x} \frac{dx}{dt} + \frac{\partial S_g}{\partial t} \qquad\qquad 10$$

which, upon comparison with **Equation 9**, yields the following pair of ordinary differential equations

$$\frac{dx}{dt} = \frac{q}{\phi A} \frac{df_g}{dS_g} \qquad\qquad 11$$

and

$$\frac{dS_g}{dt} = 0 \qquad\qquad 12$$

Equation **12** represents the characteristic line along which **Equation 11** must be solved. Thus, the left-hand side of **Equation 11**, otherwise known as the frontal advance formula, denotes the rate of travel of a plane surface* of any given saturation S_g; df_g/dS_g then refers to the point where the saturation is S_g.

The model is now completely formulated and only requires a solution of **Equations 8** and **11**. However, since analytical relationships between relative permeabilities and fluid saturations are not generally available, the problem is best solved graphically.

Now it happens that the curve of f_g versus S_g is often **S**-shaped[+] when capillary pressure is neglected (see Figure 3.2a), so that the saturation distribution (Figure 3.2b) is also **S**-shaped or multivalued. This led Buckley and Leverett to the assumption that the saturation gradient can at most become infinite; that is, it cannot reverse itself.

Thus on the basis of conservation of materials, they drew the vertical line bc, Figure 3.2b, so that the shaded areas would be equal. The saturation distribution at any given time is then given by the curve abcd. The point be where the saturation changes abruptly is called the front, ahead of which the saturation distribution is still the original.

Since **Equation 1** represents the fundamental conservation form of problems involving shock waves, it has been suggested[3] that this sharp saturation front is in line with a shock solution of **Equation 11**. The shock condition apparently arises from the neglect of dissipative forces such as capillarity as was shown by Fayers and Sheldon[10].

Now, when the initial saturation distribution is such that the gas saturation is at or below its critical value, Welge[41] devised a simple method for determining the saturation at the front. By material balance, the cumulative volume of injected or invaded gas, $Q = \int q \, dt$ must equal the pore volume behind the front times the change in S_g, that is

$$Q = \phi \, A \, \Delta x \, (\bar{S}_g - S_{gi}) \qquad\qquad 13$$

where $\bar{S}_g$ is the average gas saturation behind the front.

Substitution in **Equation 11** gives

$$\frac{df_g}{dS_{gf}} = \frac{1}{\bar{S}_g - S_{gi}} \qquad\qquad 14$$

The subscript f refers to conditions at the front.

---

*In cylindrical flow the "rate of travel" is not dr/dt but dr^2/dt, or in general, the rate of increase of volume behind the surface of constant saturation.

[+]The frontal curve approximates the mathematical logistic curve for a single cycle of growth, empirically represented by $y = a_0/ (1 + \exp(-a_1 (x - x_0)))$; a_0, a_1 are constants to be determined and x_0 is a reference value.

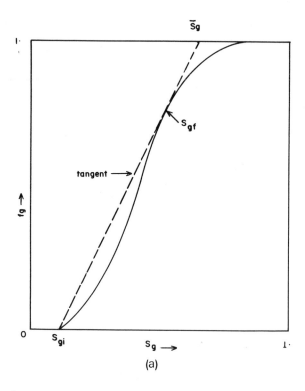

(a)

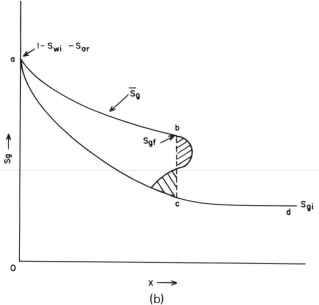

(b)

Figure 3.2. Fractional flow and saturation distribution.

From geometry, $\bar{S}_g$ is given by the intercept of the tangent from the initial gas saturation S_{gi} on the $f_g = 1$ line. (See Figure 3.2a).

The average saturation $\bar{S}_g$ behind any saturation S_g higher than S_{gf} (that is, behind the front) is

$$\bar{S}_g = \frac{1}{x}\int_0^x S_g \, dx = \frac{1}{x}\left(S_g x + \int_{1-S_{or}-S_{wi}}^{S_g} x \, dS_g\right) \tag{15}$$

Substituting $\dfrac{Q}{\phi A}\dfrac{df_g}{dS_g}$ for x in the integral and changing limits of integration

gives

$$\bar{S}_g = S_g + \frac{1-f_g}{df_g/dS_g} \;, \; S_g > S_{gf} \tag{16}$$

From geometry, $\bar{S}_g$ is given by the intercept on the $f_g = 1$ line of the tangent to f_g, at S_g. This is the Welge tangent method. It might be added that **Equation 16** is also valid for radial flow, if $\bar{S}_g$ is the volumetric average.

If the initial gas saturation is greater than the critical value, it can be shown by preservation of material balance (see Appendix B) that the Welge tangent should still be drawn from the initial condition (S_{gi}, f_{gi}), in order to obtain the saturation at the front S_{gf}. However, since the f_g values plotted are incorrect and are not known in the frontal range because of unknown capillary forces, this has led to the suggestion[31] that the tangent should be drawn from the origin. As yet there has been no experimental evidence of either conjecture although numerical studies by Gottfried[15] for water-oil displacement seem to verify the former initial condition theory.

From an intuitive point of view however, the higher the initial gas saturation in the reservoir, the lesser are the chances of forming a displacement front. The tangent technique using initial conditions seems more to bear out this fact and at the same time establishes the saturation at the inflexion point on the fractional flow curve, as the maximum value of the displacing phase at which a front would form.

In practice however, the fractional flow curve usually has a steep gradient because of the highly unfavorable viscosity ratio in gas injection operations, and the gas saturation at the front may vary insignificantly by either method.

Some interesting engineering information is also readily available from the B-L theory. Let us integrate the frontal advance **Equation 11** over the entire length L of the reservoir. We then get

$$\int_0^L dx = \frac{\int_o^t q \, dt}{\phi A} \cdot \frac{df_g}{dS_g} \tag{17}$$

or

$$Q_i = \frac{\int_o^t q \, dt}{\phi A L} = \left(\frac{df_g}{dS_g}\right)^{-1} \;, \; S_g \geqslant S_{gf} \tag{18}$$

Thus, the cumulative pore volumes of injected gas, Q_i, required to attain a particular gas saturation $S_g \geqslant S_{gf}$ at the producing well is determined by simply taking the reciprocal of the slope of the fractional flow curve at S_g, corresponding to the saturation at the well.

By the same token, the producing gas-oil ratio when Q_i pore volumes of gas have been injected is easily shown to be

$$R = \frac{f_g B_o}{(1 - f_g)B_g} + R_s \qquad\qquad 19$$

If, however, gas dissolves or comes out of solution, because of lack of equilibrium with the injected gas, **Equation 18** may be corrected[22] for the dissolved gas. Let ΔR_s be the amount of gas dissolved (negative if released), then the amount Q_i, at reservoir conditions, injected to breakthrough is

$$\bar{S}_g - S_{gi} + (1 - \bar{S}_g - S_{wi})\frac{\Delta R_s B_g}{B_o} = \left(\frac{d f_g}{d S_g}\right)^{-1} \qquad\qquad 20$$

where the right-hand term corresponds to the tangent from $\bar{S}_g$ on the $f_g = 1$ line. $\bar{S}_g$ is obtained by trial and error so that both sides of the equation are satisfied, using the f_g versus S_g plot. A direct tangent method of obtaining the frontal gas saturation in the presence of undersaturated oils is given in Pirson[31].

In conclusion, it is important to recall that the B-L displacement theory assumes no interphase mass transfer and that pressure variations are not great over most of the volume of the system. When major phase changes are involved, as in high pressure condensing or vaporizing gas drives, special computation techniques are required. These are discussed in the next chapter.

It should also be pointed out that because of the unfavorable viscosity ratios involved in gas-oil displacements, the upper part of the fractional flow curve breaks sharply indicating that there is not much oil recovery after breakthrough. Further more, gas-oil ratios may be governed more by the permeability profile and areal sweep efficiencies than by saturation-distance distributions. In that case, the B-L theory should be used only to determine breakthrough recoveries. After all, calculations after breakthrough are highly theoretical, particularly that of invaded gas.

3.2 Measuring Relative Permeability Ratios

Equations 11 and **16**, section 3.1, have been the basis of a laboratory method[20] of determining relative permeability ratios, particularly k_g/k_o, with air or nitrogen as the displacing phase. The average saturation in the core is known from the production; df_g/dS_g from **Equation 17**, section 3.1, L being the core length and $\int qdt$ the cumulative production; and f_g from the gas-oil ratio, **Equation 19**, section 3.1. Then the saturation near the end of the core is found from **Equation 16**, section 3.1, and at this saturation $k_g/k_o = R$ μ_g/μ_o. The pressure drop must be small; $\int qdt$ is corrected to mean pressure and the gas-oil ratio R, to the down-stream pressure. The velocity must be sufficient to minimize capillary-end or frontal zone effects and the core long enough to make these and the initial inlet transient effects of minor importance.

It is presumed that, while the saturation at the very outlet end may be governed by capillarity, the saturation close to the end will be given by **Equation 16**, section 3.1, and the gas-oil ratio here will be essentially the producing gas-oil ratio. This method of measuring relative permeabilities has been found[7] to be more reliable than direct measurements.

3.3 Decline Curve Analysis

Because of the linear relation between S_g and ln (k_g/k_o) over a considerable range somewhat after breakthrough, a very useful production decline relationship can be developed[27] which provides for a simple appraisal of gas injection projects.

An approximate empirical relationship between relative permeability ratio and oil saturation is

$$ln \ (k_g/k_o) \ \sim \ - \ aS_o \qquad\qquad 1$$

If the solution gas be neglected, the gas-oil ratio formula **Equation 16**, section 2.3, becomes

$$k_g/k_o \ = \ \frac{\mu_g \ B_g \ q_g}{\mu_o \ B_o \ q_o} \ \sim \ b \ q_g/q_o \qquad\qquad 2$$

where a and b are empirical constants.

Substituting **Equation 2** into **Equation 1** and differentiating with respect to time we get

$$a \ \frac{dS_o}{dt} \ = \ \frac{d \ (ln \ q_o/q_g)}{dt} \qquad\qquad 3$$

Also, from the results of **Equation 8**, section 2.1, we can write

$$a \ \frac{dS_o}{dt} \ = \ - \ \frac{a \ B_o \ q_o}{A \ h \ \phi} \qquad\qquad 4$$

Now, for a constant gas injection rate q_g, **Equation 4** is substituted into **Equation 3** and the resultant differential equation is integrated to give

$$q_o \sim \frac{A\ h\ \phi}{a\ B_o\ t} \qquad\qquad 5$$

Thus, after breakthrough, the oil production rate declines as the reciprocal of time. Similarly, it can be shown that the cumulative oil production is approximately a linear function of the logarithm of the cumulative gas through-put. Also, the producing gas-oil ratio increases exponentially with the cumulative oil recovery.

3.4 Gravity Segregation

With regard to the gravity segregation process the corresponding fractional flow formula, without capillary effects, follows from **Equation 8**, section 3.1. Thus

$$f_g = \frac{1 - \dfrac{A\ \lambda_o}{q}\ \Delta\ \rho_{o\text{-}g}\ g\ \text{Sin}\ \alpha}{1 + \dfrac{\lambda_o}{\lambda_g}} \qquad\qquad 1$$

which relates the gravitational forces and the mobility ratio in the system.

The gravitational forces are seen to be a function of the formation dip, the density difference between the oil and injected gas and more importantly, since it is the only externally controlled parameter, the injection (or production) rate. Thus the gravity segregation mechanism is rate sensitive, shifting the fractional flow curve to the right (see Figure 3.3), which implies higher recoveries, as the injection (and offtake) rate is decreased. Likewise, it can be appreciated that since capillary forces tend to oppose gravitational effects, then the more depleted are the pores, the slower will be the rate of drainage.

On the other hand, B-L calculations assume the formation of a vertical front. Both capillary and gravitational forces cause a distention in the front thereby affecting breakthrough recovery.

Example 5

This problem is a continuation of Example 4 of Chapter 2. It is required to determine if an economical oil production rate of 5500 STB/D will give significant gravity segregation. The reservoir is repressured to 3500 psia.

Data:

S_g:	0.1	0.2	0.25	0.3	0.35	0.40	0.45
F_g:	0.060	0.448	0.700	0.833	0.921	0.953	0.976
f_g:	0.0114	0.228	0.443	0.603	0.735	0.816	0.880

Where, in field units

$$f_g = F_g \left\{ 1 - \frac{1.13\,\lambda_o\,A\,\Delta\rho_{o\text{-}g}\,\text{Sin}\,\alpha}{144\,q} \right\} \qquad 2$$

and

$$F_g = \left\{ 1 + \frac{\lambda_o}{\lambda_g} \right\}^{-1} \qquad 3$$

From previous data at 3500 psia (Example 4): $\mu_o = 0.486$ cps; $\mu_g = 0.0239$ cps; $B_o = 1.565$; $B_g = 0.000670$ RB/SCF; $A = 150,000$ ft.^{2}; $\rho_o = 41.4$ #/ft^3; $\rho_g = 15.7$ #/ft^3.

Solution:

$q_o = 5500 \times 1.565 = 8610$ RB/D

From the graph of F_g versus S_g, Figure 3.3, the average residual gas saturation without gravity is $\bar{S}_g = 0.355$. With gravity and at the production rate of 8610 RB/D, the f_g versus S_g curve gives $\bar{S}_g = 0.475$.

Let us estimate whether a few wells can maintain the above rate as up-dip wells are shut-in. The productivity of a well assuming 100 psi drawdown is

$$q_o = \frac{7.07\,k_o\,h\,\Delta p}{\mu_o\,\ln \dfrac{r_e}{r_w}} = \frac{7.07\,(1.0)\,(15)\,(100)}{0.486\,(2.3)\,\log \dfrac{1000}{0.25}} = 2600 \text{ RB/D}$$

assuming a "perfect" completion for the r_w and r_e indicated.

What are the expected recoveries by depletion and by pressure maintenance?

Oil recovery by depletion (from Example 4) $= \dfrac{17.5 + 1.9}{66.7} = 29\%$

Oil recovery by pressure maintenance $= \dfrac{32 + 1.9}{66.7} = 51\%$

If injection is stopped after production of 28 MM STB (pressure still at 3500 psia), the saturations are

$$S_o = \frac{(66.7 - 29.9)\,1.565}{119} = 0.484$$

$$S_g = 1 - 0.10 - 0.484 = 0.416$$

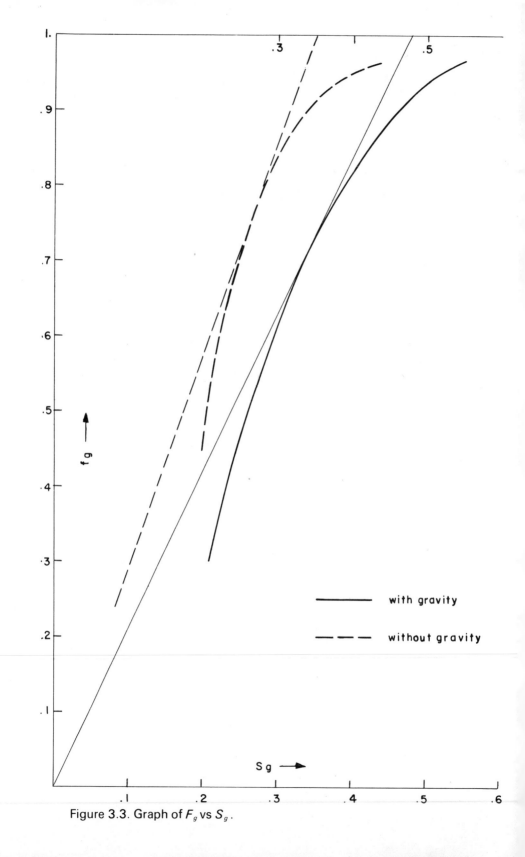

Figure 3.3. Graph of F_g vs S_g.

Using interpolation or empirical equations for R_s, B_o and R_s (see section 2.6), it is found that the pressure would drop to 1900 psia after production of the last 4 MMSTB. The corresponding saturations are then

$$S_o = \frac{32.8\,(1.357)}{119} = 0.374$$

$$S_g = 0.526$$

From the Welge plot at $S_g = 0.526$, $f_g = 0.80$ and the gas-oil ratio

$$R = \left(\frac{0.80}{1-0.80}\right)\left(\frac{1.357}{0.001235}\right) + 580 = 5000 \text{ SCF/STB}$$

The slope of the tangent is 1.5, therefore the total through-put is,

$$Q_i = 1/1.5 = 0.67 \text{ pore volume, or } \frac{0.67}{0.526} = 1.27 \text{ displacement pore volumes.}$$

3.5 Modifications for Counterflow [3][8]

As we have already seen, the treatment of the displacement problem under gravity segregation is directly handled by the fractional flow rate formula, **Equation 1**, section 3.4, which applies, in the case of upstructure gas injection, to the downward movement of all fluids (gas and oil). However, when counter-flow occurs, this results in negative fractional flow values, values greater than unity and when oil and gas rates are equal and opposite, in infinite values since $u_t = 0$.

Therefore one modification required of the B-L theory for counterflow is the use of actual instead of fractional flow rates.

By simple algebraic manipulation of the Darcy flow **Equations 2 and 3**, section 3.1, and the capillary pressure relationship **Equation 4**, section 3.1, we obtain the actual flow rate formulas for oil and gas.

$$u_o = \left\{\frac{1}{\lambda_g} + \frac{1}{\lambda_o}\right\}^{-1}\left\{\frac{u_t}{\lambda_g} + \frac{dP_c}{dx} - g\,\Delta\rho_{o\text{-}g}\,\text{Sin}\,\alpha\right\} \qquad 1$$

$$u_g = \left\{\frac{1}{\lambda_g} + \frac{1}{\lambda_o}\right\}^{-1}\left\{\frac{u_t}{\lambda_o} - \frac{dP_c}{dx} + g\,\Delta\rho_{o\text{-}g}\,\text{Sin}\,\alpha\right\}$$

along a stratum in the x-direction making a positive angle with the horizontal.

It is apparent that these equations differ from the fractional flow formulas by the factor u_t. If $u_t = 0$, the right hand brackets are equal and of opposite signs.

Also, if λ_o differs considerably from λ_g, the counterflow rate is largely governed by the smaller mobility λ_o, as postulated by Pirson.[31]

The corresponding continuity equations

$$\frac{\partial u_g}{\partial x} = - \phi \frac{\partial S_g}{\partial t} \qquad\qquad\qquad 2$$

$$\frac{\partial u_o}{\partial x} = - \phi \frac{\partial S_o}{\partial t}$$

lead to the analogous frontal advance equation

$$\frac{dx}{dt} = \frac{1}{\phi} \frac{\partial u_g}{\partial S_g} = \frac{1}{\phi} \frac{\partial u_o}{\partial S_o} \qquad\qquad 3$$

where dx/dt is the velocity of some saturation S_g and the derivatives are those corresponding to that saturation in the system at the particular time at which the velocity is to be calculated.

It should be emphasized, however, that although **Equations 1** and **2** are valid for arbitrary formation dip, in effect, gravity always acts vertically giving rise to a complicated situation wherein a saturation gradient develops between the top and bottom of the stratum. For this reason we will restrict our discussion to flow in a vertical reservoir.

Now that the counterflow problem has been formulated in an analogous manner with the B-L theory, we may conjecture that the Welge tangent rules apply. It is of interest to show the B-L type of calculation as modified for some hypothetical cases of counterflow.

The prediction of vertical saturation distribution as a function of time for counterflow, starting with an initial distribution, requires some modification of the methods commonly used for horizontal or vertical displacement without counterflow. Specifically, some assumption regarding the initial saturation distribution must be made, such that the frontal advance **Equation 3** can be applied over the entire saturation range within which relative permeabilities are not zero.

If the initial saturation is more or less uniform, Sheldon et al[36] suggests assuming that the top and bottom of the system go immediately to residual saturations with respect to the oil and gas phases, respectively. Also, allowance must be made for the fact that two "fronts" moving toward each other (if formed) cannot pass each other; that multivalued saturations or gradients are physically meaningless and that the final distribution must preserve over-all material balance.

Let us first consider the special case of a reservoir with a uniform initial saturation distribution and in which $u_t = 0$. The plot of u_o or u_g against saturation is bell-shaped (Figure 3.4a) and not a function of time if capillarity is

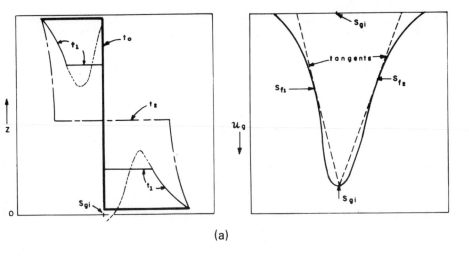

(a)

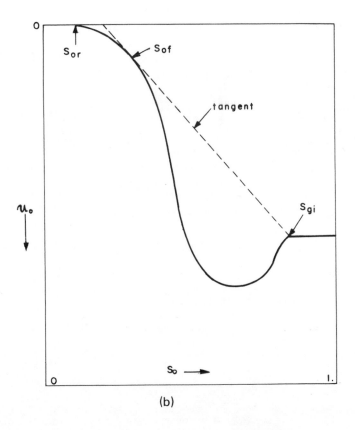

(b)

Figure 3.4. The tangent method with gravity counterflow.

neglected. A downward (oil front) and an upward (gas front) moving fronts are formed at t_1, which, of course, cannot pass each other. The distribution at a later time t_2 is indicated. Using the Welge method, tangents are drawn from the u_g curve at the initial (uniform) saturation S_{gi}. Thus, the maximum (or minimum) saturations at the fronts, S_{f1} and S_{f2}, and average saturations behind the fronts are determined graphically in the same manner as if fractional flows were used. Similarly, the linear velocity of a front is given by $1/\phi$ times the slope $\frac{du}{dS}$ of the tangent, which is also equal to $\dfrac{u\ (S_f) - u(S_{gi})}{\phi\ (S_f - S_{gi})}$ from geometry and material balance.

Figure 3.4b shows a more common form of the plot of u_o versus S_o in which counterflow is not complete ($u_t \neq 0$). The tangent again is drawn from the initial condition which preserves material balance. However, the second tangent cannot be drawn; hence there is only one front.

Experimental evidence seems to indicate that both the Darcy equations and the modified Buckley-Leverett-Welge method are valid for counterflow. The main difficulty, however, stems from two sources. In the first place, during the segregation process the upper part of the reservoir is governed by a drainage process while the lower part is controlled by imbibition. Thus, capillary pressures and relative permeabilities require complex relationships. Finally, vertical permeability variations must also be taken into account.

Example 6

A reservoir had produced 6 MMSTB of $38°$ API ($Y_o = 0.835$) oil and 6 MMMSCF of gas in five years, when the natural water drive hit the upper row of wells. It is proposed to recover part of the "attic oil" trapped at the top by injecting gas in the upper row of producing wells (or some of them), in the hope that an economical amount of this oil can be recovered by counterflow.

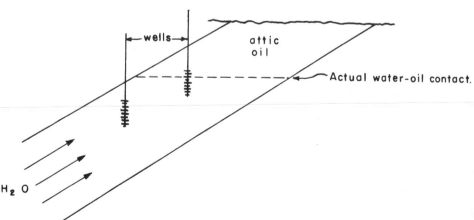

Figure 3.5. Attic oil trapped by the natural water drive.

Data:

ϕ	= 0.27	$p_1 =$ 2500 psia		$p_{act} =$ 2000 psia	
S_{wi}	= 0.20	R_s:	.800	.670 SCF/STB	
$\bar{k}$	= 1 darcy	B_o :	1.320	1.270 RB/STB	
α	= 16°	μ_o:	0.385	0.420 cp	
h_v	= 20 ft.	μ_g:	0.02	0.016 cp	

γ_g = 0.7 (air = 1); $\bar{Z}$ = 0.80 for injected and formation gas at T = 660° R. Average width of reservoir at upper row of wells = 3000 ft. At p = 3000 psia, μ_o = 0.35 cp. and μ_g = 0.024 cp. Estimated remaining volume of reservoir oil above the upper row of wells (attic oil) = 2 MMSTB.

Calculations: Assume a pressure around the injection wells of 3000 psia as necessary to force back the water drive and force liberated gas back into solution. From a plot of f_g versus S_g at p = 3000 psi, where

$$f_g = \left(1 + \frac{\lambda_o}{\lambda_g}\right)^{-1} \qquad\qquad 4$$

the Welge tangent method gives the average saturation in the injection bubble as $\bar{S}_g$ = 0.32. Note that the fractional flow formula does not include gravity effects since movement is both up and down. The data used in **Equation 4** is as follows:

S_g:	0.1	0.2	0.3	0.4	0.5	0.6
k_o/k_g:	133.	10.	1.5	0.5	0.06	0.0067
f_g:	0.099	0.593	0.905	0.965		

At $\bar{S}_g$ = 0.32, the relative permeabilities of the oil and gas phases are calculated using Corey *et al's* formulas, **Equation 8**, section 2.6.

$$K_{no} = \left(\frac{S_o}{1 - S_{wi}}\right)^4 = 0.13$$

and

$$K_{ng} = \left\{1 - \left(\frac{S_o}{1 - S_{wi}}\right)^2\right\}\left\{1 - \frac{S_o}{1 - S_{wi}}\right\}^2 = 0.102$$

Calculate ρ_o and ρ_g at p = 2500 psia since "banks" were formed and the gas at p = 3000 psia did not penetrate through the banks, but previously liberated gas was redissolved.

$$\rho_o = \frac{62.4\,(5.615)\,\gamma_o + 28.8\,Y_g R_s/379}{5.615\,B_o} = 45.2\ \#/\text{ft.}^3$$

$$\rho_g = \frac{28.8\,\gamma_g}{379\,(5.615)\,B_g} = 9.0\ \#/\text{ft.}^3$$

The counterflow formula, in field units, is

$$q_o = 0.00783\ KA\ \Delta\rho_{o\text{-}g}\ \text{Sin}\ \alpha\ \left(\frac{\mu_o}{K_{r_o}} + \frac{\mu_g}{K_{rg}}\right)^{-1},\ \text{RB/D} \qquad 5$$

where

$$A = 3000\,(20)\,\cos 16° = 57700\ \text{ft.}^2$$

then

$$q_o = 0.00783\,(1)(57700)(36.2)\ \text{Sin}\ 16°\ \left(\frac{0.385}{0.13} + \frac{0.02}{0.102}\right)^{-1} = 1430\ \text{RB/D}$$

This value may be conservative, as more favorable saturations may build up. From **Equation 5**, calculate other values of q_o at different saturations and estimate the maximum rate of counterflow, namely $q_{o\,\text{max}} = 2300$ RB/D at $S_g = 0.20$.

In order to recover 60% of the attic oil (2 MMSTB) at a counterflow rate of $q_o = 1430/1.320 = 1080$ STB/D would require 1110 days shut in time plus production time, say three years.

Assume injection of enough gas to give the calculated displacing gas saturation, $\bar{S}_g = 0.32$, in the attic pore volume 2 MMSTB x 1.320/0.8 = 3.3 MMRB, so as to make room for the counterflow. At 3000 psia and reservoir temperature, the amount of gas needed is

(pore volume) $\bar{S}_g/B_g = 3.3\,(0.32)/(0.89 \times 10^{-3}) = 1.2$ MMMSCF

Actually this volume of gas must be increased somewhat to account for a desired recession of the invaded water.

3.6 The Dietz Model

This fluid displacement model, although first published in 1953[8], remained fairly unknown until its full potential was demonstrated by Hawthorne[16] (1960) in a series of interesting experiments. We will use Hawthorne's development of the theory in this discussion.

In contrast to Buckley-Leverett's theory, the Dietz model is a two-dimensional representation of the movement of the fluid interface wherein the injected gas invades the oil reservoir as a tongue overriding the oil. Moreover, the oil and gas are assumed to flow separately, the oil behind the gas-oil contact being at its residual value (see Figure 3.6).

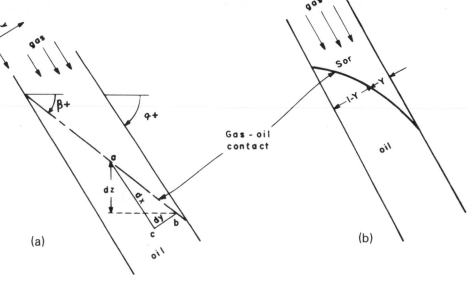

Figure 3.6. Geometry of the Dietz Model.

Both theories make the common assumption that a sharp interface divides the gas-oil contact. While this latter assumption is satisfactory in the B-L model for thin sands where the capillary forces tend to dampen the formation of fingers, it would seem to limit the validity of Dietz theory to thick sands. Nevertheless, Outmans[30] has shown that the Dietz theory is equally valid for fingering in thin sands.

With the above assumptions in mind, the development of the model follows from the concept of potential across the interface shown in Figure 3.6a.

Darcy's law in potential form along the x-axis, is

$$u_o = - \rho_o \lambda_o \frac{\partial \Phi_o}{\partial x}$$

$$u_g = - \rho_g \lambda_g \frac{\partial \Phi_g}{\partial x}$$

1

where the potential of an element of fluid is defined by

$$\Phi = \frac{dp}{\rho} + gz$$

2

z being positive upwards. Since the fluids are considered incompressible Equation 2 can be integrated to give

$$\Phi = \frac{p}{\rho} + gz = \text{constant}$$

3

If we further assume that the interface is very long relative to the (perpendicular) thickness of the formation, such that $\Phi_b \sim \Phi_c$, then the potential difference along the interface ab for both the oil and gas phases is given by

$$p_{oa} - p_{ob} + \rho_o \, g \, dz = \left(\frac{u_o}{\lambda_o}\right)_{a-c} dx \qquad\qquad 4$$

and

$$p_{ga} - p_{gb} + \rho_g \, gdz = \left(\frac{u}{\lambda_g}\right)_{a-c} dx \qquad\qquad 5$$

If on the interface the capillary pressure is constant, **Equation 4** becomes

$$p_{ga} - p_{gb} + \rho_o \, gdz = \frac{u_o}{\lambda_o} \, dx \qquad\qquad 6$$

Subtracting **Equation 6** from **Equation 5**, gives

$$g \, \Delta \rho_{o-g} \, dz = \left(\frac{u_o}{\lambda_o} - \frac{u_g}{\lambda_g}\right) dx \qquad\qquad 7$$

From the geometry of Figure 3.6a.

$$\frac{dz}{dx} = \frac{\sin \beta}{\cos (\alpha - \beta)} \qquad\qquad 8$$

Hence **Equation 7** can be written as

$$\frac{\sin \beta}{\cos (\alpha - \beta)} = \frac{\left(\dfrac{u_o}{\lambda_o} - \dfrac{u_g}{\lambda_g}\right)}{g \, \Delta \rho_{o-g}} \qquad\qquad 9$$

If the interface is stable, $u_o = u_g = u$ constant along the streamlines which are parallel to the formation, and **Equation 9** reduces to

$$\frac{\sin \beta}{\cos (\alpha - \beta)} = \frac{u \left(\dfrac{1}{\lambda_o} - \dfrac{1}{\lambda_g}\right)}{g \, \Delta \rho_{o-g}} \qquad\qquad 10$$

Which may be solved explicit for the tilt of the interface

$$\tan \beta = \frac{u \left(\dfrac{1}{\lambda_o} - \dfrac{1}{\lambda_g}\right) \cos \alpha}{g \, \Delta \rho_{o-g} - u \left(\dfrac{1}{\lambda_o} - \dfrac{1}{\lambda_g}\right) \sin \alpha} \qquad\qquad 11$$

For low dipping formations, **Equation 11** simplifies to

$$t_{an} \ \beta \ = \ \frac{u \left(\dfrac{1}{\lambda_o} \ - \ \dfrac{1}{\lambda_g} \right)}{g \ \Delta \ \rho_{o \text{-} g}} \qquad\qquad 12$$

Note that if the interface is not stabilized, **Equation 9** applies.

For uniform linear separate flow, Darcy equations may be written as

$$u_o \ = \ - \ \rho_o \ \lambda_o \ (1 - Y) \ \frac{\partial \ \Phi_o}{\partial x} \qquad\qquad 13$$

and

$$u_g \ = \ - \ \rho_g \ \lambda_g \ Y \ \frac{\partial \ \Phi_g}{\partial x} \qquad\qquad 14$$

where $(1 - Y)$ is the normalized (fractional) cross-sectional area of oil flow and Y, the fraction exposed to gas flow as shown in Figure 3.6b. With these modifications **Equation 9** is easily rearranged to give

$$f_g \ = \ \frac{1 - \dfrac{\lambda_o \ A \ (1 - Y)}{q} \ g \ \Delta \ \rho_{o \text{-} g} \ \dfrac{Sin \ \beta}{Cos \ (\alpha - \beta)}}{1 + \dfrac{\lambda_o \ (1 - Y)}{\lambda_g \ Y}} \qquad\qquad 15$$

Note that when the interface is parallel to the formation dip ($\alpha = \beta$), **Equation 15** reduces to the equivalent B-L **Equation 1**, section 3.4. This emphasizes the fact that B-L disregards the interface formation and only considers flow along a streamline. In other words, it is valid only in thin formations. Also, relative permeabilities in **Equation 15** are based on fluid distributions whereas the B-L formula refers to a given saturation. Thus, the B-L frontal advance **Equation 11**, section 3.1, may also be used in conjunction with **Equation 15** if the B-L saturations are defined as

$$S_g = Y (1 - S_{ni} - S_{or}) \qquad\qquad 16$$

and

$$S_o = (1 - Y) (1 - S_{wi}) + Y S_{or} \qquad\qquad 17$$

where S_{or} is the residual oil saturation in the gas invaded zone.

In conclusion, the Dietz model which postulates no oil flow behind the front should be a good approximation for displacement of gas or oil by water. Although the displacement of oil by gas would not be expected to fit the Dietz assumptions closely, Hawthorne's experimental studies showed that the theory adequately describes such systems thereby indicating little movement of oil behind the gas front. In fact, this would seem to confirm the idea that oil production after gas breakthrough is mainly attributable to the stripping action of the gas.

3.7 Interface Stability

Of paramount importance in two-fluid displacement is the stability of the interface. Small perturbations at this surface may cause fingering of the displacing phase and consequently an inefficient displacement mechanism.

We now write **Equation 10**, section 3.6, in the following form

$$\text{Sin } \alpha - \frac{dy}{dx} \text{ Cos } \alpha = \frac{u \left(\frac{1}{\lambda_o} - \frac{1}{\lambda_g} \right)}{g \, \Delta \, \rho_{o \text{-} g}} \tag{1}$$

and observe that the interface becomes parallel to the formation when $\frac{dy}{dx} = 0$. This then represents the condition of instability of the interface and **Equation 1** becomes

$$u_c = \left| \frac{g \, \Delta \, \rho_{o \text{-} g} \, \text{Sin } \alpha}{\frac{1}{\lambda_o} - \frac{1}{\lambda_g}} \right| \tag{2}$$

where u_c is referred to as the critical velocity at which instability (fingering) occurs.

Equation 2, however, does not specify whether stability is achieved above or below this critical value. In order to clarify this aspect of the problem we need to analyze the role of the specific parameters in **Equation 12**, section 3.6. In fact, we should use **Equation 11**, section 3.6, but in practice the value of the gravitational term in the denominator dominates over the corresponding viscous term. Note also that cos α is always positive.

Consider, for example, the case of updip gas injection in an oil reservoir. Then from physical considerations

$$\lambda_g > \lambda_o$$

and

$$\Delta \, \rho_{o \text{-} g} > 0 \tag{3}$$

Therefore **Equation 12**, section 3.2, can be written symbolically as

$$\tan \beta \rightarrow \frac{u \left(\begin{array}{c} \text{positive} \\ \text{term} \end{array} \right)}{\left(\begin{array}{c} \text{positive} \\ \text{term} \end{array} \right)} \tag{4}$$

Now, in order to avoid overriding of the oil by the injected gas, we want the interface angle β to be positive small (refer to Figure 3.7a) and therefore conclude that the critical velocity as determined by **Equation 2**, must be a maximum, that is, $u_c = u_{max}$.

We now analyze the less frequent case of downdip gas injection in an oil reservoir, for example, the attic oil problem. Here the same conditions **Equation 3** prevail, β must be kept positive small (Figure 3.7b), and therefore $u_c = u_{max}$.

As a final example, let us look at the case of down dip water injection in a dry gas reservoir. **Equation 12**, section 3.6, is modified by replacing oil terms with water terms.

Now, from physical considerations, we have

$$\lambda_g > \lambda_w$$

5

$$\Delta \rho_{w \cdot g} > 0$$

and therefore

$$\tan \beta \rightarrow \frac{u \text{ (positive)}}{\text{(positive)}}$$

6

However, as shown in Figure 3.7c, underriding by the injected water causes the tilt angle β to grow negatively. Therefore the critical velocity must be a minimum, $u_c = u_{min}$. This result indicates that dry gas and light oil reservoirs with strong edge water drives should be produced at the highest rates possible in order to minimize water fingering. Similar geometrical and physical analyses must be made for other reservoir/injection fluid combinations.

As a final practical observation, it has been found that the tilt angle of the gas injection front can be predicted using Dietz method, if the injection rate is less than the critical rate given by **Equation 2**. Similarly the B-L model describes accurately the displacement of oil by gas drive and gravity drainage when flow rates are less than one-half the critical rate. It should also be pointed out that the Dietz model has nothing to do with distortion of interfaces due to wells, such as "coning" or "fingering."

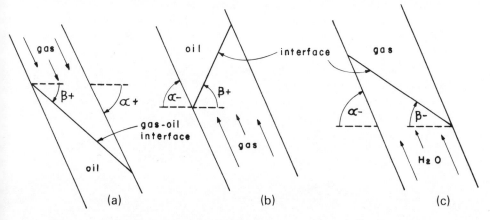

Figure 3.7. Gas injection interface angle β.

3.8 Horizontal Sweep Efficiency

Horizontal or areal sweep efficiency may be defined as the fraction of the total area of the oil reservoir that is invaded by the injection fluid as it advances from injection to producing wells. The relative area swept when the displacing fluid first reaches the producing wells is referred to as breakthrough sweep efficiency and that after the appearance of the displacing phase, as after-breakthrough sweep efficiency. Its value will evidently be in the range of zero to unity, being greater after breakthrough than at breakthrough.

The concept of breakthrough sweep efficiency constitutes mainly an economic criterion since it reflects the oil recovery prior to the beginning of surface handling of the displacing fluid. In gas injection projects, however, the production of high volumes of injected gas does not involve costly operational problems as, for instance, in the case of water injection.

Gas injection operations differ in another way from those of water injection in that the former are generally developed without regard to well pattern regularity and with many more producing than injection wells. This, in part, is due to the fact that gas injection operations are conducted more for pressure maintenance purposes than as a secondary recovery process. Also, the injectivity of the wells tend to be higher for gas injection.

There are several methods available for determining sweep efficiency. These fall into three broad categories: (1) analytical, (2) numerical, and (3) experimental. Analytical techniques are of little value in irregular well patterns and therefore will not be discussed any further.

Numerical methods are obviously the most flexible since they can be self-sufficient, thereby requiring the use of a computer, or they can be used in conjunction with experimental results. The most popular self-sufficient method is that due to Higgins and Leighton[17] and modified by LeBlanc and Caudle[24]. It has the added advantage that the computation also gives the well injectivity as a function of time which is important for equipment design. The method, however, is valid only for systems with stable injection-resident fluid interfaces. In other words, the mobility ratio cannot be much larger than unity.

The mixed numerical-experimental methods are the most useful within the framework of the main theme of this book, namely, to describe a methodology which is amenable to desk calculations. In effect, the streamline-potential distribution of the reservoir system is obtained experimentally (field plotter, etc.) and the advance of a particle along several streamlines is calculated. The resultant relative position of each particle gives the advance of the injection front.

Mathematically, the time required for a fluid particle to move to any position along a streamline s is

$$t = \int_o^s \frac{ds}{u_s} = \frac{\phi\mu}{k} \int_o^s \frac{ds}{\left|\dfrac{dp}{ds}\right|} \qquad\qquad 1$$

Equation 1 is integrated graphically from a plot of ds/dp versus s along the entire length of the fastest travel time streamline. The area so obtained is proportional to the time required for breakthrough of the injected fluid. Similar graphs of *ds/dp* versus *s* are prepared for other streamlines in the system. By superimposing the area obtained for the fastest travel time streamline on the subsequent graphs, yields the relative position of each particle on its individual flow path and hence the injection fluid front.

Experimental methods for measuring sweep efficiency are numerous and varied, covering the range of fluid, electrical and x-ray analogs. Again, because of the irregular well-pattern restriction in gas injection operations, an electrical analog using conducting paper seems the most adequate. Such an analog, sometimes referred to as a field plotter, can be built at little cost and satisfactory accuracy.

Essentially, the wells are represented by electrodes and a wheatstone bridge provides detection of the potential distribution. Streamlines may be traced-in perpendicular to the iso-potentials or the original boundary conditions inverted and the streamlines obtained with the field plotter. Details of the method and circuit diagram are available in the literature.[32]

The streamline-potential distribution thus obtained, is used in conjunction with the numerical-graphical method explained above.

Another useful desk-type analog can be prepared with blotting paper.[12]

This ion-motion analog gives a direct graphic representation of the history of the steady movement of the front and is based on the mathematical analogy between the flow of ions in an electrolytic system and those of fluid particles in a porous medium. Other variations of analogs of the ion-motion type have been developed[2] which enable the study of systems with mobility ratios of up to 25:1 and with varying relative permeabilities throughout the reservoir.

The preceding discussion was primarily concerned with gas injection in oil reservoirs with an initial gas saturation lower than the critical saturation level. Now, in gas drives following normal depletion operations, the gas saturation is high enough so that the gas phase is completely mobile throughout the reservoir. Under these conditions the sweep efficiency at breakthrough is, in effect, 100 percent. This fact, which has been verified experimentally,[5] has the added importance that the injection-production performance of such a reservoir is identical to that of a linear system under the same conditions. In other words, the system's behavior is independent of the well geometry.

The presence of high permeability streaks in the formation gives rise to by-passing of oil by the injected gas. The by-passing may be of such extent as to make the operation unsuccessful. Recent field tests[18] seem to indicate the feasibility of selectively sealing off the high-permeability channels through the use of foam. Foams are known to possess good plugging characteristics especially when followed by gas injection.

Another well-known method of increasing sweep efficiency involves intermittent gas injection. The technique is based on the general principle that the more rapid depletion of the gas content and pressure in the highly permeable zones during an injection recess will induce a cross-flow into them from the tighter strata. On resumption of gas injection, the oil will be swept out from the partly resaturated permeable layers more rapidly and efficiently than if it had to be driven out directly by gas invasion into the tight strata while by-passing continues in the permeable zones.

Sweep efficiency is notoriously low in viscous-oil reservoirs because the displacing fluid, gas or water, tends to finger. Recently, a novel approach to an old idea of well pattern rotation seems to offer some promise. The method[11] combines early pattern rotation with gas injection prior to water-flooding.

Essentially, gas is injected into the reservoir until a short time after breakthrough, so as to build up a gas saturation over a reasonably large area around the producing wells. The injection pattern is rotated; that is, producing wells become injection wells and vice-versa, and the reservoir then water-flooded. Lab tests indicated 20 to 30 percent more oil recovery than by plain water injection, using an oil of 550 centipoises.

In theory, the fingers developed by the gas serve as channels for the injected water when the injection wells are alternated. The role of the injected gas is simply that of creating a gas saturation in the reservoir which apparently favors waterflooding. In contrast, simple pattern rotation, as practiced long ago, increases sweep efficiency by changing the sweep pattern.

There exists laboratory evidence[23] that prior gas injection increases water-flood displacement efficiency (although field observations are not conclusive), probably because of the presence of many other influencing parameters acting simultaneously.

Other operational techniques and injection fluid combinations that have been used to improve sweep efficiency are discussed in the following chapter, since these methods are mainly associated with miscible type displacement.

3.9 Vertical Sweep Efficiency

The third and final efficiency factor to be considered in fluid injection operations, is the so-called vertical sweep efficiency. It is a measure of the advance of the injection fluid profile, originated by the presence of inhomogeneities in the fluid and rock properties throughout the formation.

Over the years, several methods for computing this efficiency factor have been developed. Schoeppel[34] gives an excellent chronological review of the evolution of these methods.

The basic reservoir model consists of a linear layered-cake analog, with uniform rock and fluid properties in each layer, arranged in descending order of absolute permeability. The fluid displacement mechanism can either be the Buckley-Leverett model or one of separate flow, oil ahead and gas behind the front. If a non-linear model is desired, the Le Blanc-Caudle[24] technique (section 3.8) may be used in each layer.

We will describe the Dykstra-Parsons-Johnson treatment of the oil displacement problem mainly because its semiempirical approach allows for quick computations. Also, the method is widely used and been shown to give reliable results in field cases[37,39]. Although the method was originally developed using water-oil displacement data, the resultant correlations can be interpreted for gas-oil displacement calculations.

Basic data obtained from some 200 flood tests performed on about 40 California core samples were correlated by Dykstra and Parsons[9], using four fundamental variables. These are: V, vertical permeability variation; λ, mobility ratio; S_{wi}, initail water saturation; and, R_o, fractional recovery of oil in place at a given producing water-oil ratio (WOR). Their original graphical correlation was further simplified by Johnson.[19]

When a step-wise regression analysis of Johnson's graph was made, it was generally found that the most critical parameter in determining oil recovery is the permeability variation V. The other three independent variables, WOR, λ and S_{wi}, showed an almost equal influence on recovery, but their correlation coefficients were on the average six times smaller than that of V. Thus, the following simplified relationship

$$R_o = 0.47 - 0.39\, V,\ 0 \leqslant V \leqslant 1.0 \tag{1}$$

which has a satisfactory 12 percent standard error of estimate, provides an easy approximation of oil recovery. Interestingly also, **Equation 1** would seem to statistically verify the Naar-Henderson[28] imbibition model which establishes a maximum recovery efficiency of 50 percent.

The parameter V is a statistical measure of the absolute permeability profile of the reservoir and is related to the standard deviation σ_k by[39]

$$\sigma_k = ln\left(\frac{1}{1-V}\right) \tag{2}$$

and to the Lorenz coefficient, L_c, by

$$L_c = erf\left(\frac{1}{1-V}\right) \tag{3}$$

where *erf* refers to the error function, defined in the next chapter.

V is best obtained from core analysis data. The measured absolute permeability k is arranged in descending order, grouping any repeated values. The logarithm of k is then plotted on a probability graph, from which V is defined as

$$V = \frac{k_{50\%} - k_{84.1\%}}{k_{50\%}} \tag{4}$$

In most reservoirs, however, core data is relatively scarce on a reservoir basis so that the engineer is faced with the problem of establishing the number of discrete layers, each of uniform properties, into which the reservoir is to be zoned. Craig[4] recently presented a tabular guide for selecting the minimum number of equal-thickness layers necessary to describe the reservoir. In relative terms, his results indicate that for gas injection in a typical reservoir (V = 0.6) similar performance results are obtained if the reservoir were assumed to be divided into 100 or 20 layers.

The main value of Craig's guide is in saving computer time. However, the basic problem still remains: how to specify the discrete layers since the heterogeneous profile varies from well to well. In addition, the engineer is interested in calculating the changes in oil and gas rates as a function of time. This latter task of computing performance behavior for each layer and then taking sums at any one time to obtain the total production from a well, is rather cumbersome.

A more convenient approach to this problem would be to consider the reservoir as an infinitesimal layered-cake.[29] Thus, rather than taking a number of layers of finite thickness, we assume the layers to be of infinitesimal thickness, with continuously increasing or decreasing permeability.

If q_{unit} (kt) is the fluid production rate for a given permeability k, per foot of sand, at some time t, then the total production rate is given by

$$q \text{ (total)} = \int_{o}^{h} q_{unit} \ (kt) \ dz \qquad\qquad 5$$

where z is the height in feet corresponding to k and h is the total formation thickness.

In general, a universal curve of q_{unit} versus kt may be calculated using some specific displacement model. The integration represented in **Equation 5** can be done graphically by plotting q for a given permeability against the height corresponding to that permeability and taking the area under the curve from z = 0 to h. For machine computation, of course, equivalent numerical methods must be used. The method is illustrated in Example 7.

All of the layered-cake models used in determining vertical sweep efficiency are based on the assumption of no cross-flow between adjacent layers. Both stochastic[39] and deterministic[14] approaches have been used to study the effects of cross-flow on sweep efficiency. The problem is complex in that cross-flow is affected by both differential capillary and viscous gradients within the system which in turn are affected by the saturation distribution of the fluids.

Of course, cross-flow also depends on the ratio of the vertical to horizontal permeabilities.

We may generalize the results of the various investigations on the topic in the following way. It appears that for unfavorable mobility ratios, as occur in gas drives, cross-flow promotes instability at the front thereby reducing recovery efficiency. Thus, the neglect of cross-flow in such a truly stratified system would generate an optimistic behavior of the system. By the same token, to consider the reservoir uniform would grossly overestimate the performance.

In the final analysis of gas drive projects, Muskat[27] has well pointed out that from a purely economic point of view, reservoir uniformity is not so exacting a requirement for successful gas injection as it is for waterflooding. This is so because although heterogeneity brings on early breakthrough of the injection fluid, the additional cost of handling excessive water production is far greater than that associated with high gas-oil ratios.

Example 7

Calculate the total oil production rate for the permeability profile shown in Figure 3.8. Assume the universal production rate history shown in Figure 3.9.

Solution: The permeability profile, Figure 3.8, shows a range from 10 to 90 millidarcies, with some weighting in the middle region. From Figure 3.9, oil production begins at $kt = 75$ md — months, so for the highest permeability, 90 md, this is $75/90 = 0.83$ months. There is, therefore, no production earlier than this. The calculation is illustrated below in tabular form for two, four and 10 months.

The total production rate at any time is obtained from the area of Figure 3.10. These are 440, 150 and 12 for two, four, and 10 months, respectively. From the total rates calculated at a number of different times a curve of rate as a function of time is obtained (Figure 3.11). Calculation of gas rates, using the gas curve of Figure 3.9, is entirely similar.

If porosities, initial saturations and relative permeability relations are different for all permeabilities, then, in general, separate universal oil and gas curves similar to those of figure 3.9 have to be constructed for each of the several permeabilities used.

Calculations:

		2 months			4 months			10 months		
k	z+	kt	q/kz++	q/z	kt	q/kz++	q/z	kt	q/kz++	q/z
90	12	180	0.25	22.5	360	0.02	1.8	900	0	0
70	10	140	.95	66.5	280	.06	4.2	700	0	0
50	6	100	.97	48.5	200	.16	8.0	500	0	0
30	2	60	0	0	120	1.04	31.2	300	0.04	1.2
10	0	20	0	0	40	0	0	100	.97	9.7

+ from Figure 3.8
++ from Figure 3.9

(text continues on page 88)

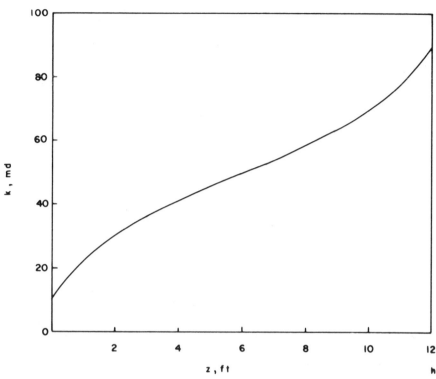

Figure 3.8. Permeability profile assumed for production calculations.

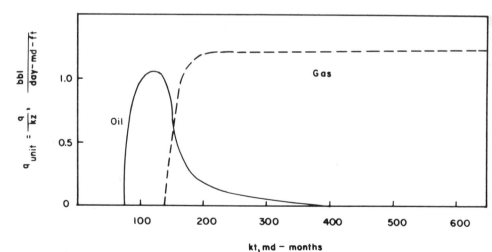

Figure 3.9. Assumed production rate history for one foot of 1 millidarcy sand.

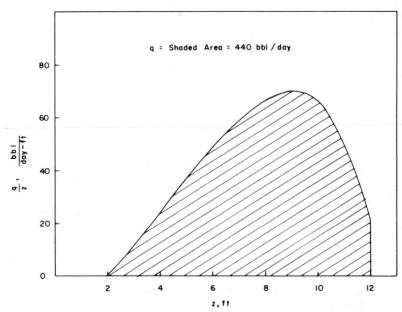

Figure 3.10. Graphical integration of rate over the formation thickness at two months.

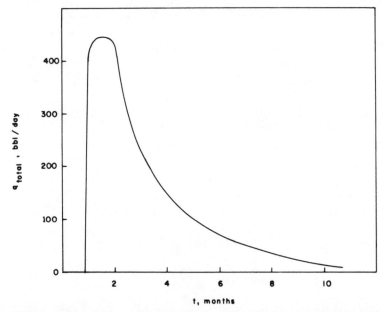

Figure 3.11. Calculated oil production rate for permeability profile of Figure 2.8.

3.10 Mobility Ratio

Throughout the discussions in this chapter it has become fairly evident the important role that the mobility ratio plays in determining displacement, areal and vertical sweep efficiencies of fluid injection projects. As a matter of practice it is now well established to define mobility ratio λ as the ratio of the displacing phase mobility $\lambda_g = k_g/\mu_g$, to that of the displaced phase $\lambda_o = k_o/\mu_g$, specific reference being made here to the gas injection model problem.

We note that the above definition of relative mobility ratio involves four parameters or the product of two ratios: relative permeability and relative viscosity. The viscosity ratio μ_o/μ_g is a function of the pressure at which the injection project is being conducted and is essentially constant. The value of this ratio in typical gas injection projects is large, normally greater than 100 even for medium gravity oils.

Defining the relative permeability ratio is a more complex question however. As will be recalled from the assumptions inherent in the Dietz model, the relative permeability ratio is there defined as a constant, being measured at the saturation of the corresponding phases in each separate flow region. In other words we have $k_g (S_{or})/k_o (S_{gi})$.

While the Dietz model would theoretically appear to be more representative of a water displacement process wherein the oil saturation in the water invaded zone is normally low, this is not necessarily the case with gas displacement.

In fact, we can conjecture that an adequate definition of relative permeability ratio, for any immiscible fluid displacement process, should include the saturation distribution of the displacing phase behind the injected fluid front. Such a distribution is plainly evident from the Buckley-Leverett model and this is the direction taken in the following theoretical development originally due to Warren and Cosgrove.[40]

Consider a flow process in a one-dimensional porous medium with simultaneous parallel flow of incompressible and immiscible fluids. The continuity **Equation 11**, section 3.1, can be written as

$$\int_o^x A (x)\, dx = \frac{\int_o^t q\, dt}{\phi} \frac{df_g}{dS_g} \qquad\qquad 1$$

or

$$\Psi (S_g) = \frac{Q(t)}{\phi} \frac{df_g}{dS_g} \qquad\qquad 2$$

where $Q(t) = \int_o^t q\, dt$ is the cumulative volume of gas injected.

Since

$$\Psi(S_g) = \int_o^x A(x)\, dx \tag{3}$$

is a monotone increasing function of x, then for a given value of $Q(t)$

$$x = x(S_g) \tag{4}$$

Thus, differentiation of **Equation 2** with respect to x gives

$$A(x) = \frac{Q(t)\, d^2 f_g}{\phi \; dS_g^2} \; \frac{\partial S_g}{\partial x} \tag{5}$$

To determine the effective mobility of the displacing phase, consider the form of Darcy's equation which describes the total volumetric flow rate

$$q(t) = \frac{\lambda_g A(x)}{f_g} \; \frac{dp}{dx} \tag{6}$$

Equation 6 integrates to

$$\Delta p = \frac{q(t)\, \mu_g}{k} \int_o^x \frac{f_g\, dx}{k_{rg} A(x)} \tag{7}$$

If we assume a fairly uniform saturation behind the front, we may represent **Equation 7** as follows

$$\Delta p \sim \frac{q(t)\, \mu_g \bar{f}_g}{k_g} \int_o^x \frac{dx}{A(x)} \tag{8}$$

Thus, the effective mobility behind the displacement front is given by

$$\lambda_g = \frac{\bar{K}_g}{\mu_g f_g} = \frac{\int_o^x d_x}{\mu_g \int_o^x \frac{f_g\, dx}{K_g A(x)}} \tag{9}$$

Ahead of the displacement front, the saturation (S_{gi}) and fractional flow (f_{gi}) of the displacing phase are uniform. Consequently

$$\lambda_o = \frac{k_o(S_{gi})}{\mu_o(1 - f_{gi})} \tag{10}$$

Then, the effective mobility ratio, $\lambda = \lambda_g/\lambda_o$, can be obtained from the ratio of **Equations 9** and **10**. Thus

$$\lambda = \frac{\mu_o\,(1 - f_{gi})}{\mu_g\,k_o\,(S_{gi})} \cdot \frac{\int_o^x dx}{\int_o^x \dfrac{f_g\,dx}{k_g\,A\,(x)}} \qquad 11$$

Using **Equation 5**, and changing the variable of integration, **Equation 11** becomes

$$\lambda = \frac{\mu_o\,(1 - f_{gi})}{\mu_g K_o\,(S_{gi})} \cdot \frac{\displaystyle\int_{S_{g1}}^{S_{g2}} \frac{f_g{}^{11}}{A^2\,(S_g)}\,d S_g}{\displaystyle\int_{S_{gi}}^{S_{g2}} \frac{f_g\,f_g{}^{11}}{k_g\,A^2(S_g)}\,d S_g} \qquad 12$$

where the double prime denotes the second derivative with respect to saturation.

Warren and Cosgrove evaluated **Equation 12** for systems of different geometry and for gas and water drives. The type of displacement mechanism was defined using the pertinent relative permeability relationship, Corey *et al.'s* approximation for drainage processes (gas drive) and Naar-Henderson's equations (see section 2.6) for imbibition processes (water drive). Part of their results are shown in Figures 3.12 and 3.13 for linear ($A\,(x)$ = constant) and radial ($A\,(x)$ = $a + b\,(x)$) systems, before and after breakthrough. Other geometric systems fall between these two limits and can always be construed in terms of these two basic geometries.

These results clearly show, on the one hand, a definite geometric effect on mobility ratio. Also, even for very adverse viscosity ratios (~ 100) as are typical of gas drives, we observe that the mobility ratio attains values less than 7. After breakthrough, however, the mobility ratio increases appreciably (Figure 3.13).

Craig *et al.*[5] had previously suggested a simple method of estimating the displacing fluid mobility by using the average saturation value obtained by the Welge tangent technique. For the sake of comparison, the mobility ratio values calculated by Craig *et al.'s* method are also indicated in Figure 3.12.

It is immediately apparent that the average saturation method slightly overestimates the effective mobility ratio for viscosity ratios usually encountered in gas drives. The difference occurs because at high viscosity ratios there is a steep saturation gradient behind the displacement front and the use of the average saturation, which implies a zero gradient, no longer represents an adequate approximation. It is likewise evident that the Dietz definition of mobility ratio will yield values even much higher than Craig *et al.'s*.

In conclusion, the results of this section indicate that effective mobility ratios greater than ten will rarely be encountered in gas-injection operations and that the displacement mechanism is inherently stable. This also explains the high success of displacement models which are mainly designed for systems with mobility ratios not too much greater than unity and without consideration of the fingering phenomenon.   *(text continues on page 92)*

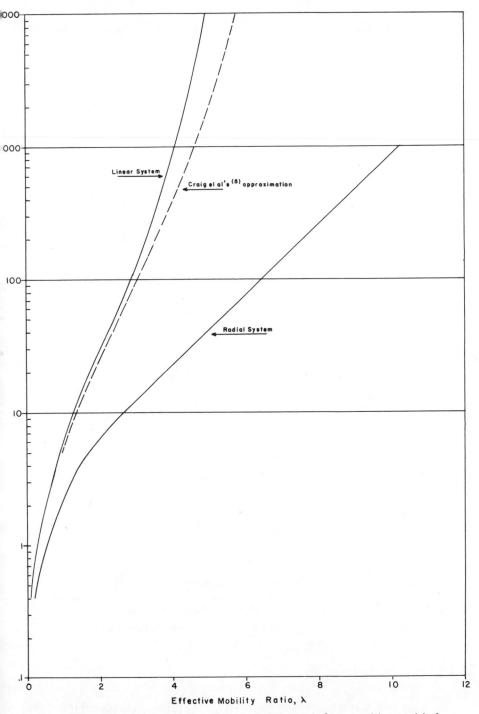

Figure 3.12. Effective mobility ratio at breakthrough for gas drives with f_{gi} = S_{gi} = 0 (adapted from Warren and Cosgrove[40]).

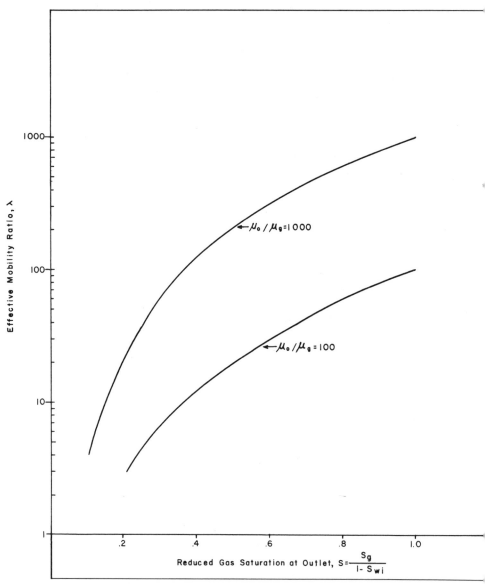

Figure 3.13. Effective mobility ratio after breakthrough for gas drives in a linear system with $f_{gi} = S_{gi} = 0$ (adapted from Warren and Cosgrove[40]).

3.11 Scale-Up Methods

A rigorous theoretical treatment of many flow problems in porous media is as yet beyond the scope of analytical methods. In some cases, even when mathematical solutions are available, it is difficult to interpret the results in a physical context as usually occurs when the solution is not in closed form. The behavior of such problems, however, can be studied by means of scaled flow model experiments.

Scale groups prove very useful also in data correlation of both field and laboratory data. They permit a homogenization of apparently disperse information. An example in point is the Leverett J-function which consolidates capillary pressure data obtained from cores of varying porosities and permeabilities. The Leverett scale group is defined as

$$J(S_g) = P_c \left\{ \sigma \sqrt{\frac{\phi}{k}} \cos \theta \right\}^{-1} \qquad 1$$

where σ is the interfacial tension between the fluids saturating the porous medium and θ is the wetting angle between the sand grains and the fluids. The other variables have already been defined.

There are essentially two general methods for obtaining scale groups for a specific problem. These methods are denoted dimensional analysis and inspectional analysis. Dimensional analysis is more fundamental in its appraoch in that it assumes a complete knowledge of all variables affecting the problem. These variables are then combined by algebraic techniques into independent dimensionless groups. Inspectional analysis takes the view point that the equations governing the problem are known and from these, the corresponding scale groups are obtained.

Of course, both methods give the same groups if all the controlling parameters are represented in the equations formulated. This latter condition, however, is not necessary all of the time since physical considerations may justify the elimination of some variables *a priori*. Therefore, in practice, inspectional analysis usually yields fewer groups than dimensional analysis. Since the equations governing the gas drive model, the analysis of which is the main goal of this section, are well established, the inspectional analysis method will be used to obtain the corresponding scale groups.

Inspectional analysis can be approached from either of two angles. One method requires making the variables in the controlling differential or integral equations, dimensionless. This is accomplished by redefining each variable in terms of some reference parameter of the problem, such as, length of the system, residence time, injection flux or some other known criterion. This technique is usually more difficult in that the definition of the dimensionless variables requires some physical insight and experience with the problem.

The second method merely requires three simple steps:[21]

1. Verify that the governing equations are dimensionally homogeneous.
This means that every term in the physical equations must have the same

dimensional formula in terms of, for example, the fundamental quantities: m, mass; L, length; and t, time. It is therefore necessary that dimensional constants, such as the gravitational constant, be considered.

2. Reduce each equation to a generalized dimensional form, omitting differential signs and numerical constants.

3. Divide through by any one term to render all the terms dimensionless.

These are the required independent dimensionless or scale groups. The procedure will be demonstrated with the well-known diffusivity equation for flow of an undersaturated fluid through a porus medium.

For a linear system, the pertinent form of the equation is

$$\frac{\partial^2 p}{\partial x^2} = \frac{1}{\eta} \frac{\partial p}{\partial t} \qquad\qquad 2$$

We must first verify its dimensional homogeneity by expressing each term of **Equation 2** in fundamental units, namely

$$\frac{m/L\,t^2}{L^2} = \frac{1}{L^2/t} \frac{m/L\,t^2}{t} \qquad\qquad 3$$

It is seen at once that every term has the dimensions $[m/L^3\ t^2]$ and therefore the original **Equation 2** is dimensionally homogeneous.

The second step requires reducing **Equation 2** to generalized dimensional form. We write

$$\frac{p}{L^2} = \frac{1}{\eta} \frac{p}{t} \qquad\qquad 4$$

$$\text{I} \qquad\quad \text{II}$$

whereupon dividing through by I, we obtain the only dimensionless group, $L^2/\eta t$.

As will be recalled, this similarity group has been used to transform **Equation 2** into an ordinary differential equation, thereby simplifying its solution. This variable transformation is sometimes referred to as the Boltzmann transformation or the method of combination of variables. The transformation is only valid if the initial and boundary conditions can be expressed exclusively in terms of the new variable.[35] The knowledge of the scale group also allows us to completely define the behavior of the problem with a single plot of p versus $x^2/\eta t$.[33]

Now that we have demonstrated the general methodology for obtaining scale groups from the governing equations, we will analyze the gas injection problem. The controlling set of equations (see section 2.1) for a linear reservoir is

$$\frac{\partial}{\partial x} \left\{ \lambda_o \frac{\partial}{\partial x} \left(p_o + \rho_o \, gz \right) \right\} = \phi \, \frac{\partial S_o}{\partial t} \qquad\qquad 5$$

$$\frac{\partial}{\partial x} \left\{ \lambda_g \frac{\partial}{\partial x} \left(p_g + \rho_g \, gz \right) \right\} = \phi \, \frac{\partial S_g}{\partial t} \qquad\qquad 6$$

$$p_c = p_g - p_o = J(S_g) \, \sigma \, \sqrt{\frac{\phi}{k}} \, \cos \theta \qquad\qquad 7$$

$$S_o + S_g + S_{wi} = 1 \qquad\qquad 8$$

It can be easily shown that these equations are dimensionally homogeneous. We therefore proceed to express them in generalized dimensional form. **Equation 5** becomes

$$\frac{1}{L} \left\{ \frac{\lambda_o}{L} \, \left(p_o + \rho_o \, gL \, \mathrm{Sin} \, \alpha \right) \right\} = \frac{1}{t} \qquad\qquad 9$$

or

$$\frac{\lambda_o \, p_o}{L^2} + \frac{\lambda_o \, \rho_o \, g \, \mathrm{Sin} \, \alpha}{L} = \frac{u}{t} \qquad\qquad 10$$

where the time dimension was replaced by the definition of velocity, namely $t = L/u$. This was done mainly for convenience since most injection projects are conducted under constant injection rate conditions which would be represented by u in this case.

Similarly, **Equation 6** becomes

$$\frac{\lambda_g \, p_g}{L^2} + \frac{\lambda_g \, \rho_g \, g \, \mathrm{Sin} \, \alpha}{L} = \frac{u}{L} \qquad\qquad 11$$

Likewise, **Equation 7** is expressed as

$$\frac{uL}{\lambda_g} - \frac{uL}{\lambda_o} = \sqrt{\frac{\phi}{k}} \; \sigma \cos \theta$$

$$\underset{\text{I}}{} \quad \underset{\text{II}}{} \qquad \underset{\text{III}}{}$$

12

where the pressures of the oil and gas phases were replaced by the corresponding Darcy group or equation ($u = \dfrac{\lambda_o p_o}{L} = \dfrac{\lambda_g p_g}{L}$).

Introducing the Darcy group in **Equations 10** and **11**, we obtain

$$\frac{u}{L} + \frac{\lambda_o \, \rho_o \, g \, \text{Sin} \, \alpha}{L} = \frac{u}{L}$$

$$\underset{\text{IV}}{} \quad \underset{\text{V}}{} \qquad \underset{\text{VI}}{}$$

13

and

$$\frac{u}{L} + \frac{\lambda_g \, \rho_g \, g \, \text{Sin} \, \alpha}{L} = \frac{u}{L}$$

$$\underset{\text{VII}}{} \quad \underset{\text{VIII}}{} \qquad \underset{\text{IX}}{}$$

14

Note that the saturation **Equation 8** is not utilized because saturation is by definition dimensionless. For this same reason dimensionless constants such as porosity and formation volume factors are not included. ϕ is retained in **Equation 12** because it is here considered as a single variable, $\sqrt{\dfrac{\phi}{k}}$.

We may now form the various dimensionless groups using **Equations 12, 13** and **14**. Dividing through **Equation 12** by term I yields two groups, namely

$$\frac{\lambda_g}{\lambda_o} \; , \quad \frac{k_{rg} \, \sqrt{\phi k} \; \sigma \cos \theta}{u \, \mu_g \, L}$$

15

Similarly, dividing through **Equations 13** and **14** by the u/L term yields two additional groups

$$\frac{\lambda_o \, \rho_o \, g \, \text{Sin} \, \alpha}{u} \; , \quad \frac{\lambda_g \, \rho_g \, g \, \text{Sin} \, \alpha}{u}$$

16

These two latter groups can be more conveniently combined into the single group

$$\frac{\lambda_g \, \Delta \rho_{o \, - \, g} \, g \, \text{Sin} \, \alpha}{u}$$

17

Although the correct procedure for combining groups requires division of one into the other, if this were done for **Equation 16** the newly formed group would be of the form

$$\frac{\lambda_g \, \rho_g}{\lambda_o \, \rho_o} \qquad\qquad 18$$

which essentially contains the mobility ratio group already present in **Equation 15**. Moreover, the ratio ρ_g/ρ_o is less sensitive than the difference $\Delta\rho_{o\text{-}g}$. Hence the 'justification' for creating the group **Equation 17**.

Combining the scale groups previously derived, the gas injection problem is defined by the following parameters

$$\lambda = f\left(\underbrace{\frac{\lambda_g \, \Delta\rho_{o\text{-}g} \, g \, \text{Sin} \, \alpha}{u}}_{\text{Froude group}}, \underbrace{\frac{k_{rg} \sqrt{\phi \, k} \, \sigma \cos\theta}{u \, \mu_g \, L}}_{\text{Leverett group}}\right) \qquad 19$$

$$\underbrace{}_{\text{Mobility ratio}}$$

Thus, a graph, say, of mobility ratio versus the Leverett group, with the Froude group as parameter, would completely define the gas-oil displacement problem. The literature contains a considerable number of correlations of this type.[6,25]

The Leverett group is the ratio of the capillary-interfacial forces to the viscous forces while the Froude group reflects the relative importance of the gravitational and viscous forces. The third group, mobility ratio, in a sense, may be considered the most important of the three groups, since under varying experimental or operational conditions both the Froude and Leverett groups may be insignificant. For instance in a horizontal reservoir the Froude group is zero.

Furthermore, it is interesting to note that both the mobility ratio and Froude group enter explicity in the Buckley-Leverett and Dietz models (see sections 3.1 and 3.6).

The extension of the above method to the multidimensional case is straightforward and the details are left to the reader. The resulting groups for the three dimensional isotropic reservoir, are

$$\lambda = f\left(\frac{L_1}{L_2}, \frac{L_1}{L_3}, \frac{\lambda_g \, \Delta\rho_{o\text{-}g} \, g \, L_3}{u \, L_1}, \frac{\lambda_g \sqrt{\dfrac{\phi}{k}} \, \sigma \cos\theta}{u \, L_1}\right) \qquad 20$$

where L_1, L_2, L_3 refer to the lengths of the system in the x, y, and z directions, respectively.

If the reservoir is anisotropic $(k_x \neq k_y \neq k_z)$ where the k's are the principal values of the absolute permeabilities, x, y, z being the principal directions, the equivalent isotropic permeability has been shown[26] to be the geometric mean, $(k_x \cdot k_y \cdot k_z)^{1/3}$. The length dimensions must also be distorted according to the relations

$$x = \widetilde{x}\,(k_x/k_o)^{\frac{1}{2}}; \quad y = \widetilde{y}\,(k_y/k_o)^{\frac{1}{2}}; \quad z = \widetilde{z}\,(k_z/k_o)^{\frac{1}{2}} \qquad 21$$

with k_o an arbitrary value of k and $\widetilde{x}, \widetilde{y}, \widetilde{z}$ are the new coordinates. For instance, in a two dimensional system the ratio of the lengths of the system, L_1 and L_2, follows from **Equation 21**.

$$L_1 : L_2 = \sqrt{k_x} : \sqrt{k_y} \qquad 22$$

These modifications can easily be incorporated into the scale groups **Equation. 20**.

It should be emphasized that this geometric distortion of the model is not limited to permeability but can be validly extended to the broader concept of anisotropic hydraulic diffusivity or transmissibility. An excellent discussion of the theory of scaling ratios in petroleum reservoirs is given by Geertsma, Croes and Schwarz.[13]

Finally, it is important that we realize that these scaling criteria presuppose three fundamental conditions: (1) identical relative permeability curves for the model and prototype, (2) a linear relationship between model and prototype of the capillary pressure curve, and (3) similar boundary conditions. These restrictions are rather stringent and constitute a major obstacle in a complete scaling of field problems.

3.N Closure

This chapter was largely concerned with an examination of the different "simple" models which arise in connection with determination of the dynamic behavior of immiscible fluid displacement processes in porous media.

Consideration is first given to a detailed discusssion of the two fundamental mechanistic models of displacement theory: Buckley-Leverett and Dietz. The differences, similarities and particular areas of application are emphasized. In particular, the Dietz model is shown to provide an adequate criterion for determining both the angle of tilt and stability of the injection-resident fluid interface. The necessary modifications of the Buckley-Leverett-Welge method to examine the problem of gravity segregation with counterflow are also developed, describing the technique with an example of recovery of attic oil.

Following the treatment of displacement models, some methods for computing horizontal and vertical sweep efficiencies are explained. In regard to the Dykstra-Parsons-Johnson experimental method, results of a multiple regression analysis are given which indicate the strong dependence of recovery efficiency on the heterogeneity of the reservoir.

Next, a theoretical development of the measure of mobility ratio in two-fluid displacement systems, is presented. The object is to show that gas-oil systems are inherently stable which provides a justification for the use of performance models which do not consider the fingering phenomenon.

Finally, methods for obtaining scale ratios in the gas drive process are developed. The analysis is extended to anisotropic and multidimensional space. The general methodology will be used in obtaining some mathematical solutions to the miscible process in the next chapter.

Questions

1. Derive the fractional flow formula for three flowing phases in the absence of capillary and gravitational forces.

2. Assume a linear gas drive downdip in a reservoir at uniform initial gas saturation, $S_{gc} = 0.05$. Values of f_g are as follows:

S_g:	0.1	0.2	0.25	0.3	0.35	0.40	0.45
f_g:	0.060	0.448	0.700	0.833	0.921	0.953	0.976

   a. What is the gas saturation just behind the front and the average saturation in the gas invaded zone?
   b. What is the fractional oil recovery if wells are closed in as soon as injection gas appears?
   c. What is the fractional recovery if the wells are closed in at a GOR of 25000 SCF/STB? (use $B_g = 0.0005$ RB/SCF; $B_o = 1.3$ RB/STB).
   d. What is the ultimate fractional recovery for an infinite GOR?
   e. What is the total gas injected in terms of pore volumes of the invaded zone when the GOR = 25000 SCF/STB.

3. Use the Laplace transform technique to show that the solution of the convective equation

$$u \frac{\partial f}{\partial x} = \frac{\partial f}{\partial t}, \quad f(x, 0) = 0; \quad f(0, t) = 1.$$

   gives rise to a "shock" front condition or piston-type wave.

4. With regard to Example Problem 6, section 3.5, assume that 1.82 MMMSCF of gas is needed. To inject this in one year requires 5 MMSCF/D and 430 horse power. Assuming the cost of the compressor station at $400/HP; gathering and injection lines including well conversion, $50,000; and the value of the oil at $1.00/barrel after operating costs and taxes, determine whether the project is economical.

5. Analyze the case of updip water injection in a dry gas reservoir to determine if high injection rates are desirable.

6. Show that the LHS's of **Equations 10**, section 3.6, and **1**, section 3.7, are identical.

7. Show that the conservation equations

$$\text{div}\,(\rho\,\vec{u}) + \frac{\partial \rho}{\partial t} = 0$$

yield no scale information. (Note: they only help to define dimensionless variables).

Symbols (in Field Units)

a	empirical constant
A	cross-sectional area normal to flow, ft^2
b	empirical constant
B_o, B_g	formation volume factor of oil and gas, respectively, RB/STB, RB/SCF.
f_g, F_g	fractional gas flow, dimensionless
$\bar{f}_g$	average fractional gas-flow, dimensionless
f_{gi}	fractional gas flow at initial saturation conditions, dimensionless
g	gravitational constant, 1/144
h	formation thickness, ft
J	Leverett function
k	absolute permeability, md.
k_o, k_g, k_w	effective oil, gas and water permeability, md.
k_{rg}	relative permeability to gas, dimensionless
k_x, k_y, k_z	principal absolute permeabilities in x, y, z directions, md.
ln	natural log, base e
L, L_1, L_2, L_3	length measurement, ft
L_c	Lorenz coefficient, dimensionless
ln	natural log, base e
p_o, p_g	reservoir pressure of oil and gas phases, psia
P_c	capillary pressure, psia
q	volume injection rate, ft^3/day
q_o, q_g	volumetric rate of oil and gas, respectively, reservoir volume/day
q_{unit}	fluid production rate per unit formation capacity, reservoir volume/md.-ft.
Q	cumulative injected or invaded gas, ft^3
Q_i	cumulative injected or invaded gas, pore volumes

R	producing surface gas-oil ratio, SCF/STB
R_o	fractional recovery of original oil in place
R_s	solution gas-oil ratio, SCF/STB
r_e	drainage radius of a well, ft.
r_w	well radius, ft.
s	distance along a streamline, ft.
S_f	maximum saturation at the front, dimensionless
S_g, S_o	saturation of gas and oil phases, dimensionless
S_{gi}, S_{wi}	initial saturation of gas and water phases, dimensionless
$\bar{S}_g$	average gas saturation in the invaded zone, dimensionless
S_{of}, S_{gf}	oil and gas saturations at the injection front, dimensionless
S_{or}	residual oil saturation, dimensionless
t	time, days
u	oil or gas Darcy velocity, ft/day
u_c	critical oil or gas velocity, ft/day
u_o, u_g	Darcy velocity of oil and gas phases, ft/day
u_s	velocity along a streamline, ft/day
u_t	injection velocity, ft/day
V	permeability variation, dimensionless
WOR	producing water-oil ratio, RB/RB
$x, x, y, \tilde{y}$	space coordinates, ft.
Y	fractional cross-sectional area exposed to gas flow
$z, \tilde{z}$	space coordinates, ft
α	formation dip, degrees
β	interface tilt, degrees
Δ	difference operator
η	hydraulic diffusivity constant, ft^2/day
ϕ	fractional porosity, dimensionless
Φ_o, Φ_g	potential of oil and gas phases, psia
γ_o, γ_g	specific gravity of oil and gas, respectively, dimensionless
λ	mobility ratio, $\lambda = \lambda_o/\lambda_g$
λ_o, λ_g	mobility of oil and gas (= $k_o/\mu_o, k_g/\mu_g$)
μ_o, μ_g	viscosity of oil and gas, respectively, cps
ρ_o, ρ_g	density of oil and gas, #/ft^3
σ	interfacial tension, dynes/cm
σ_k	standard deviation of permeability distribution, dimensionless
Θ	wetting angle, degrees

References

[1] Buckley, S. E. and Leverett, M. C. "Mechanism of Fluid Displacement in Sands." *Trans. AIME* (1941) 146,107.

[2] Bureau, M. and Manasterski, D. "Modéle Electrolytique pour l'étude du déplacement de l'huile d'un gisement par un fluide d'injection." *Revue de l'Institut Francais du Pétrole et Annales des Combustibles Liquides,* (1963) XVIII, 369.

[3] Cardwell, W. T. "Solutions of Immiscible Displacement Equations without Triple Values." *Trans. AIME* (1959) 216, 271.

[4] Craig F. F. "Effect of Reservoir Description on Performance Predictions." *J. Pet Tech.,* October (1970), 1239.

[5] Craig, F. F., Geffen, T. M. and Morse, R. A. "Oil Recovery Performance of Pattern Gas or Water Injection Operations from Model Tests." *Trans. AIME* (1955) 204, 7.

[6] Craig, F. F. Sanderlin, J. L., Moore, D. W. and Geffen, T. M. "A Laboratory Study of Gravity Segregation in Frontal Drives." *Trans. AIME* (1957) 210, 275.

[7] Croes, G. A. and Schwarz, N. "Dimensionally Scaled Experiments and the Theories on the Water-drive Process." *Trans AIME* (1955) 204, 175.

[8] Dietz, D. N. "A Theoretical Approach to the Problem of Encroaching and Bypassing Edge Water." *Konikl Ned. Akad. Wetenschap* (1953) Proc. B56, 83.

[9] Dykstra, H. and Parsons, R. L. *The Prediction of Oil Recovery by Water Flood, Secondary Recovery of Oil in the United States,* 2nd, Ed., New York: API (1950).

[10] Fayers, F. J. and Sheldon, J. W. "The Effect of Capillary Pressure and Gravity on Two-Phase Fluid Flow in a Porous Medium." *Trans. AIME* (1959) 216, 147.

[11] Felsenthal, M. and Ferrell, H. H. *A Method for Viscous-oil Recovery,* API (1967), p. 218.

[12] Fothergill, C. A. "An Improved Blotter Model for Analog Studies." *Trans. AIME* (1957) 210, 412.

[13] Geertsma, J., Croes, G. A. and Schwarz, N. "Theory of Dimensionally Scaled Models of Petroleum Reservoirs." *Trans. AIME* (1956) 207, 118.

[14] Goddin, C. S., Craig, F. F., Wilkes, J. O. and Tek, M. R. "A Numerical Study of Waterflood Performance in a Stratified System with Cross-flow." *Trans. AIME* (1966) 237, 765.

[15] Gottfried, B. S., Guilinger, W. H. and Synder, R. W. "Numerical Solutions of the Equations for One-Dimensional Multi-Phase Flow in Porous Media." *Trans. AIME* (1966) 237, 63.

[16] Hawthorne, R. G. "Two-Phase Flow in Two-Dimensional Systems-Effects of Rate, Viscosity and Density on Fluid Displacement in Porous Media." *Trans. AIME* (1960) 219, 81.

[17] Higgins, R. W. and Leighton, A. J. "A Computer Method to Calculated Two-Phase Flow in Any Irregularly Bounded Porous Medium." Trans. (1962) 225, 679.

[18] Holm, L. W. "Foam Injection Test in the Siggins Field." *J. Pet. Tech.,* Dec. 1970, 1499.

[19] Johnson, C. E. "Prediction of Oil Recovery by Waterflood – A Simplified Graphical Treatment of the Dykstra-Parsons Method." *Trans. AIME* (1956) 207, 345.

[20] Johnson, E. F., Bossler, D. P. and Naumann, V. O. "Calculation of Relative Permeability from Displacement Experiments." *Trans. AIME* (1956) 216, 370.

[21] Johnstone, R. E. and Thring, M. W. *Pilot Plants, Models and Scale-up Methods in Chemical Engineering,* New York: McGraw-Hill, Inc., 1957.

[22] Kern, L. R. "Displacement Mechanism in Multi-Well System." *Trans. AIME* (1952) 195, 91.

[23] Kyte, J. R., Stanclift, R. J., Stephen, S. C. and Rapoport, L. A. "Mechanism of Water Flooding in the Presence of Free Gas." *Trans. AIME* (1956) 207, 215.

[24] LeBlanc, J. L. A Streamline Model for Secondary Recovery, *Soc. Pet. Eng. J.* (March, 1971) 7.

[25] Matthews, C. S. and Lefkovits, H.C.: Gravity Drainage Performance of Depletion-Type Reservoirs in the Stripper Stage, *Trans. AIME* (1956) 207, 265.

[26] Muskat, M. Flow of Homogeneous Fluids through Porous Media, Ann Arbor, Michigan: J. W. Edwards, Inc. 1946. p. 227.

[27] Muskat, M. *Physical Principles of Oil Production,* New York: McGraw-Hill Inc., 1949.

[28] Naar, J. and Henderson, J. H. "An Imbibition Model – Its Application to Flow Behavior and the Prediction of Oil Recovery." *Trans AIME* (1961) 222, 61.

[29] Nielson, R. F. "Generalized Infinitesimal Layer-Cake Theory of Reservoir Production." *Producers Monthly,* July 1966, p. 23.

[30] Outmans, H. D. "Nonlinear Theory for Frontal Stability and Viscous Fingering in Porous Media." *Trans. AIME* (1962) 225, II 165.

[31] Pirson, S. J. Oil Reservoir Engineering, New York: McGraw-Hill Inc., 1958.

[32] Sandrea, R. "Analog Simulation of Potential Fields." *Prod. Monthly,* Oct. 1967.

[33] Sandrea, R. "A Self Similar Solution of Unsteady Fow of Gases through Porous Media." *Prod. Monthly,* Feb. 1967.

[34] Schoeppel, R. J. "Waterflood Prediction Methods." *Oil and Gas J.*, Feb. 19, 1968, p. 98.

[35] Sedov, L. I. *Similarity and Dimensional Methods in Mechanics,* Academic Press, 1959.

[36] Sheldon, J. W., Zondek, B. and Cardwell, W. T. "One-Dimensional, Incompressible, Noncapillary, Two-Phase Fluid Flow in a Porous Medium." *Trans AIME* (1959) 216, 290.

[37] Snyder, R. W. and Ramey, H. J. "Application of Buckley-Leverett Displacement Theory to Noncommunicating Layered Systems." *Trans. AIME* (1967) 240, 1500.

[38] Templeton, E. E., Nielsen, R. F. and Stahl, C. D. "A Study of Gravity Counterflow Segregation." *Trans. AIME* (1962) 225, 185.

[39] Warren, J. E. and Cosgrove. J. J. "Prediction of Waterflood Behavior in a Stratified System." *Trans. AIME* (1964) 231, 149.

[40] Warren, J. E. and Cosgrove, J. J. "The Effective Mobility Ratio for Immiscible Displacement." Paper presented at 24th. Technical Conference on Production, Pennsylvania State U., Circular 66 (1963).

[41] Welge, H. J. "A Simplified Method of Computing Oil Recovery by Gas and Water Drive." *Trans. AIME* (1952) 195, 91.

4. Gas-Oil Displacement With Mass Transfer

4.0 Introduction

Up to this point we have discussed different methods, conventional material balance, and dynamic models for computing the performance of an oil reservoir under various production mechanisms. The principal assumptions usually made were homogeneous fluid and rock properties, excessive pressure and saturation gradients occupying only a small part of the total volume, and validity of the relative permeability concept to describe the basic fluid flow characteristics of the reservoir. The PVT data for the reservoir fluids is experimentally obtained assuming a differential vaporization process. Moreover, the volumes of tank oil and separator gas production are usually taken to be simply related to the oil shrinkage and to solution gas determined by the laboratory differential vaporization at reservoir temperature. We need to look into the gas liberation process of the entire system (reservoir-flow-string-surface separators) more carefully so that we may appreciate the extent and limitation of our assumptions regarding the PVT data that goes into the previously mentioned models.

It is well accepted[7] that until a critical gas saturation is attained in the reservoir, the dominant gas liberation process is that of the instantaneous or flash type since vapors formed essentially remain in contact with the liquid phase. Above the critical gas saturation, the high mobility of the gas relative to the oil gives rise primarily to a differential vaporization process (the process is not entirely differential because gas continually contacts oil ahead of it as it moves). Since this latter stage represents the major part of the reservoir's productive life, the differential vaporization is considered to define more accurately the overall liberation process.

In the flow string the gas liberation process is again flash. However, over short periods of time the process may be more adiabatic than isothermal. Finally, the liberation process in the surface separators is flash but is carried out actually at the separator temperature which is different from that of the temperature of the reservoir. To a certain extent, however, equilibrium is established because of fluid movements.

Let us examine the gas injection problem. Because of a lack of reservoir homogeneity and agitation or turbulence, only a small quantity of the injected gas may go into solution. The dry injected gas more often becomes enriched by some of the heavier components of the oil and becomes mixed with the liberated gas. This enrichment of rapidly moving gas will not produce complete equilibrium and is not represented by either the flash or differential vaporization process. The process, however, would appear (although not very convincingly!) to be more of a differential nature because of the rapid movement of the gas as it strips the remaining oil. Also, we may visualize the gas cap to be composed mainly of differentially liberated gas, the original free gas and injected gas. Again it is normally assumed that the volumetric behavior of the cap is best represented by differential data.

In the case of volatile oils — those which contain a high ($\geqslant 15$ mol percent) concentration of intermediates, C_2 through C_6 — conventional material balance methods assume that all of the free or liberated gas that enters the well remains in the gaseous phase as it is produced through the separators. In fact, the lighter the oil, the more similar are the compositions of the liberated gas and the equilibrium oil at high pressures. Because of this, additional recovery is obtained by condensation of intermediates in the solution gas that was liberated in the reservoir.

From the aforesaid, it is apparent that non-attainment of equilibrium plus the existence of a complex vaporization process, which depends more on the fluid flow characteristics of the reservoir and therefore on the production mechanism itself, may cause computational errors when the physical mechanism of the model neglects these phase effects. The importance of phase effects was long recognized in the case of condensate reservoirs but only more recently for oil reservoirs. The lag may be due to the added complexity of the latter problems and the inability to handle the large amounts of data required, without the help of a computer.

Fortunately, phase effects are less pronounced in non-volatile oils and therefore conventional material balance methods are satisfactory in these cases even when dry gas injection or gas cap expansion is the displacement process. Special modifications of the material balance method to account for phase effects have been proposed[15,22] but they tend to appreciably overestimate the reservoir performance mainly because of the assumption that the reservoir behaves as a well-stirred tank. Since this implies that phase equilibrium is reached simultaneously throughout the reservoir, the effects of vaporization or condensation are highly distorted.

For this reason we will not discuss in this chapter any of the compositional material balance methods. Nevertheless, since the reservoirs wherein phase effects are important are usually subjected to gas injection with pressure maintenance, the computational methods to be described would cover the above deficiency.

The benefits of mass transfer during gas injection operations can be further taken advantage of, to create miscible conditions and theoretically recover all of the reservoir oil. This is possible because miscibility eliminates the capillary and interfacial forces which retain substantial quantities of oil even after successful flooding under immiscible conditions. Recovery methods using miscible gas injection are also very versatile in that they can be successfully combined with other well proven methods such as waterflooding. The design of the miscible project, however, requires a high level of engineering sophistication which we will attempt to develop into a manageable procedure, after arriving at a solid understanding of the phase relationships of the mechanisms involved.

4.1 Some Thermodynamic Considerations of the Miscible Process

Miscibility may be defined as the degree of solubility of one fluid in another. All gases are completely miscible in each other but the solubility of gases in liquids depends on the chemical similarity of the fluids, the pressure and temperature of the system. Chemical similarity implies that hydrocarbon gases would dissolve more readily in hydrocarbon liquids (oils) and other organic solvents than they would, say, in connate water. Pressure affects solubiliy according to Henry's law, which states that gases dissolve in liquids more readily as the pressure of the system increases. In contrast, increasing temperatures diminish the solubility of gases in liquids.

In the case of liquid-liquid systems, the degree of miscibility depends only on chemical similarity and temperature. As two liquids are brought into contact with each other, the degree of their solubility is usually classified as (1) immiscible, or insoluble in each other, (2) partially miscible, and (3) completely miscible, implying that they dissolve in each other in all proportions, for example, liquid propane and light oils.

Ideal solution theory, analogous to ideal gas theory, is based on several assumptions such as the validity of Raoult's law of vapor pressure, conservation of volume and that the intensive fluid properties such as viscosity and density are linearly related to the relative concentration of the constituents of the

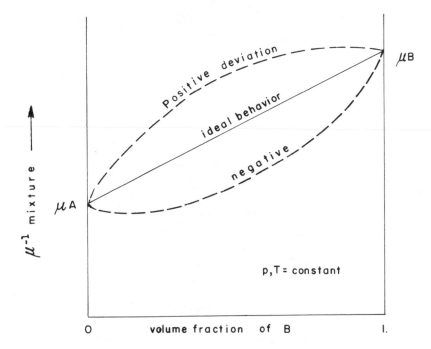

Figure 4.1. Variations from ideal solution theory in the behavior of real solutions.

solution. In real solutions, however, considerable variation from this linear behavior is observed, both positively and negatively, as shown schematically in Figure 4.1 for the solution fluidity $(1/\mu)$ of the two fluids, A and B.

Gibb's phase rule establishes the following linear relationship

$$f - c + P = 2$$

among the number of degrees of freedom f, the number of components c and the number of phases, P, present. The degrees of freedom indicate the number of excess variables (pressure, temperature and composition of each phase) that must be specified arbitrarily in order to completely characterize the state of the system. Thus, a three component system in a single phase has four degrees of freedom, namely, temperature, pressure and fraction of two of the three components. This data is normally represented on a ternary diagram at constant pressure and temperature, as shown in Figure 4.2.

Ternary diagrams have several interesting properties which must be previously mastered. Suppose component A (see Figure 4.2) is added to a mixture of B and C, represented by D. Then the composition of the 3-component system will lie

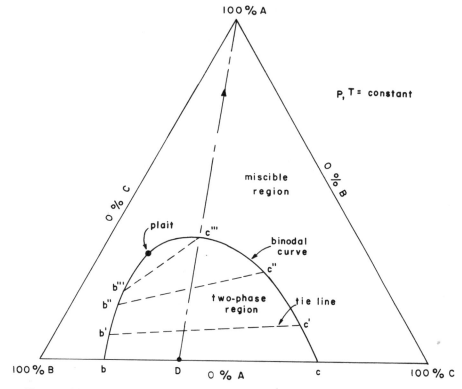

Figure 4.2. Ternary diagram of a three-component system in a single phase.

along a straight line joining A and D. This line is referred to as the composition path. Now, Figure 4.2 shows a system in which B and C are partially miscible while A is completely miscible with either B or C. The binodal curve defines the region in which mixtures of ABC are only partially miscible. Within this two-phase region mixtures of ABC would separate into two layers, a C-rich layer with compositions c, c', c'' and a B-rich layer with compositions b, b', b''. Lines joining these equilibrium compositions are termed tie lines. Thus, for instance, if a mixture D is flashed, the resulting equilibrium phase compositions are b and c assuming, of course, that the diagram is for the particular pressure and temperature of the flash process.

Tie lines are not usually parallel since A does not distribute itself uniformly between the two layers. For this same reason, the plait point or critical mixture (where the tie lines disappear) is almost never at the maximum of the binodal curve.

The slope of the tie lines is determined by the value of the equilibrium constant for the component A^*. If this constant is less than unity, the slope is downward toward the right. If it is greater than unity, the slope is upward toward the right. In most high pressure reservoirs the A component is equivalent to the so called intermediates (C_2 through C_6). The equilibrium constants for the combined intermediates will be less than one for typical reservoir temperatures and pressures.

Miscibility is accomplished by the disappearance of one of the two equilibrium layers as occurs with the C-rich layer, in this example.

Note that as A is added to D the amount of C-poor layer decreases and finally vanishes at b''', and the composition of the single phase becomes c'''. In the special case in which the line AD passes through the plait point, both phases become identical and miscibility is accomplished.

This concept of the mechanism of miscibility is of fundamental importance in that it relates directly to displacement efficiency.

It is evident that a piston-like displacement can only occur when the velocities of the displacing and displaced phases are equal, since, on the contrary, if one of the two phases lags we have a so-called leaky piston effect. We have also just seen that unless the composition path, as the fluid system moves toward miscibility, goes through the plait point, one of the two equilibrium phases tends to disappear or lag. We therefore conclude that the location of the composition path with reference to the plait point determines the effectiveness of the displacement process.

Theoretically, it is possible to control the composition path which tends to miscibility by, for example, varying the composition of fluid A, and thus exercise some control over the actual displacement mechanism. The point here is that the composition of the injected gas, say, may be varied in order to optimize displacement of a reservoir oil.

Petroleum reservoir systems are multicomponent and therefore present an obvious difficulty if one wishes to define graphically their compositional behavior. It has been verified experimentally, however, that a multicomponent hydrocarbon system may be arbitrarily represented by a three-component analog. Although, from solution theory this is not correct, the technique is useful for a mechanistic understanding of miscibility phenomena. Usually the three components used to represent these systems are: methane (C_1), intermediates ($C_2 - C_6$), and the heptanes plus (C_7^+) fraction. Evidently, this pseudo ternary diagram is rigorous if the composition of the intermediates does not change. Also, this representation of the multicomponent hydrocarbon system gives rise to the type of ternary diagram illustrated in Figure 4.2, since the intermediates are completely miscible in high pressure systems with either C_1 or

---

*Equilibrium constant is here defined as $K_A = \dfrac{\text{fraction in } C\text{-rich component}}{\text{fraction in } B\text{-rich component}}$

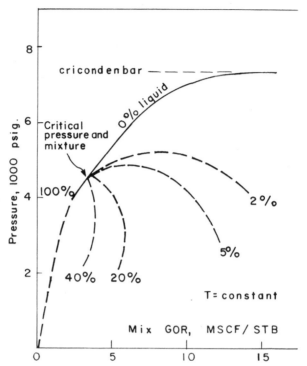

Figure 4.3. Vapor-liquid phase diagram of 5th Cockfield crude oil (after Brinkley et.al.[3]).

C_7^+, while C_1 and C_7^+ are only slightly miscible in each other. For example, C_1, is less than 0.5 weight percent soluble in decane at normal reservoir conditions.

Although the triangular diagram is valuable from a mechanistic point of view, its information must be complemented with detailed PVT and phase-behavior data on the entire system (e.g. reservoir fluids and injection gases) in order to define more clearly the engineering aspects of the problem. Normally a complete pressure-composition phase diagram (pressure versus gas-oil ratio at constant liquid lines, Figure 4.3) is prepared from which the operating conditions to maintain miscibility (single phase region) are directly established.

For example, in Figure 4.3 we see that the minimum pressure for miscibility is represented by the cricondenbar (maximum pressure) and not the critical pressure as may be suspected. It is typical of the behavior of reservoir systems that the cricondenbar may be appreciably higher than the critical pressure. The difference increases as the oil becomes heavier.

The major factor in determining a low dew point (or critical pressure) in a hydrocarbon system is the concentration of the intermediates. For example, a methane-decane mixture requires approximately 5000 psia to bring them to the critical pressure. Addition of 20 percent butane reduces the critical pressure to

3000 psia. The critical point corresponds to the plait point composition in the ternary diagram.

Theoretically, it is feasible to calculate the pressure at which a given mixture of natural gas and oil will reach a single phase using vapor-liquid equilibrium methods. However, the calculations are rather involved and require a careful analysis of the higher boiling constituents (heavier components). This single phase pressure, moreover, refers to the critical pressure and, as mentioned before, is not as important as the cricondenbar. For this reason, experimentally determined phase diagrams are mandatory.

Also, it is evident that more reliable data would be obtained using packed PVT cells, since the role of the porous medium is difficult to quantify. Needless to say, these analyses are expensive.

By the same token, because of the economics involved, it is important that the engineer investigate the feasibility of a proposed gas-injection project before extensive laboratory determinations are made. Dykstra and Mueller[9] have published a computer method of calculating fluid properties and phase composition diagrams of reservoir systems. Similarly, Grieves and Thodos[12] have presented a graphical method for calculating cricondentherm and cricondenbar pressures of multicomponent hydrocarbon mixtures of known composition. Equally important is a computer program for generating ternary diagrams for reservoir systems that was developed by Rowe and Silberberg[30].

Let us now briefly examine the concept of miscibility from a molecular point of view. Miscibility is related to mutual attraction between molecules which are unlike in type and size. This attraction also governs the tendency to condense into a liquid. The term a/V^2 in van der Waals' equation measures this force, a, being a constant for a pure compound, and the force being (roughly) inversely proportional to the fourth power of the distance between molecules. The constant a increases with molecular weight in a homologous series, as shown by the decreasing volatility of hydrocarbons with molecular weight.

One may imagine, at some pressure and temperature, a gaseous phase of 30% ethane and 70% methane in equilibrium with a liquid phase of 90% ethane and 10% methane. The higher attractive forces between ethane molecules hold down the escaping rate from the liquid, so that only 30% in the gas phase gives enough gas-to-liquid escape to balance the liquid-to-gas escape. Conversely, it takes a higher concentration of methane in the gas to balance the higher escaping tendency of methane from the liquid. The discontinuity in attractive forces at the interface gives rise to the interfacial or surface tension.

Now if ethane is pumped into the system, increasing the pressure, the increased proximity of molecules increases the attraction between ethane and methane molecules. Finally, this may become so great that all the liquid phase is "pulled" or "dissolved" into the vapor phase, this phenomenon being called "retrograde vaporization." Or, we may say "miscibility is attained." Butane and propane molecules have enough mutual attraction so that, at pressures above the vapor pressure of propane, it would be impossible, at any composition ratio, to have a second phase from and to which escaping tendencies could be balanced. In other words, complete miscibility exists. Hexane and water do not have enough mutual molecular attraction to be miscible.

4.2 Vaporizing and Condensing Gas Drives

In contrast to straightforward immiscible displacement considered in the previous chapters, two basic variations of the gas drive process, which take advantage of mass transfer effects, have been used successfully in field operations. These processes have been denoted vaporizing (equilibrium, dry gas or high pressure) gas injection and condensing (solvent, LPG, or enriched) gas injection.

In the first, or vaporizing gas drive[33], a lean gas ($\gtrsim 80$ percent C_1) is injected into the reservoir and an interchange of hydrocarbon components between the gas and resident oil takes place. At intermediate pressures, the more volatile components (intermediates) vaporize, thereby enriching the injected gas and leaving a heavier residual oil. Obviously, recovery depends on how much of the oil can be vaporized. At very high pressures, continuous movement of the gas relative to the oil, establishing new equilibrium, may result in a miscible drive which can, theoretically, recover most of the oil, as described later.

The condensing gas drive process[34], on the other hand, involves the injection of a gas which is appreciably soluble in the reservoir oil. This gas, rich in intermediates such as an LPG, would favor miscibility of the overall system and consequently all of the reservoir oil would be recovered.

We now need to complement the above qualitative explanation of the vaporizing and condensing gas drive processes with more details of the mechanisms involved.

Let us consider the pseudo ternary diagram (Figure 4.4) of a typical oil-injected gas system at reservoir conditions. ac is the dew point curve, cb the bubble point curve and c represents the critical mixture. The dotted phase boundary curve hypothetically shows the effect of injection pressure level.

Suppose that a lean separator gas D is injected into a reservoir containing an oil of composition O. The composition of the resultant mixture will follow a straight line drawn between D and O. Such a line traverses the two-phase region, thereby giving rise to two equilibrium layers each with composition indicated by the extremes of the tie line $d'o'$. Thus after infinite contacts between the injected gas and resident oil, the resultant composition of both fluids will have changed due to mass transfer to d' and o', respectively. The originally injected gas is now richer in intermediates and the reservoir oil, which was under-saturated, is now at its bubble point. This is important because the oil has essentially swelled because of solution gas, thereby reducing its viscosity.

Note that this system will never achieve miscibility. However, if the reservoir is repressured to 4000 psia, we observe that the line joining DO now falls outside of the two-phase region and misciblity is accomplished. Therefore, the tangents

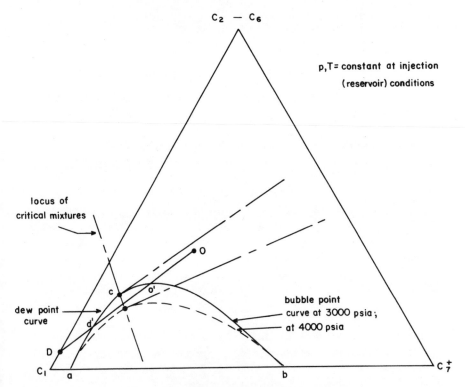

Figure 4.4. Vaporizing and condensing gas drive process.

drawn through the plait points and parallel to the limiting tie lines, define the region below which miscibility cannot be attained. In other words, reservoir oils with compositions below the tangent (e.g. viscous oils) cannot become miscible with lean injected gases except at high reservoir pressures; hence the name high pressure gas injection. Also, it is evident that miscibility with lean gases always requires that the reservoir oil be undersaturated.

We therefore conclude that the miscible vaporizing gas drive requires high injection pressures ($\gtrsim$ 3000 psia) and undersaturated oils with a fair concentration of intermediates or light oils ($\gtrsim$ 40° API). Nevertheless, even when complete miscibility is not achieved, the oil gets saturated with gas and its viscosity is reduced. In general, high API gravity, high reservoir pressure and temperature are all favorable for oil vaporization.

On the negative side, vaporization causes a severe volumetric shrinkage of the residual oil which would cause a decrease in oil-relative permeability and consequently, recovery by physical displacement would be impaired.

As we have seen so far, the requirements of the vaporizing gas drive evidently exclude its application to a wide class of reservoirs: those containing heavy oils and those at shallow depths; the latter because they cannot tolerate the high pressures involved. These limitations can be circumvented by injecting a gas rich in intermediates such as an LPG.

Assume that the ternary diagram of Figure 4.5 is an appropriate representation of the reservoir-fluid/displacing-fluid system. An enriched gas represented by the point G is to be injected into a partially depleted reservoir with a combined fluid composition O. The reservoir fluid is in the two-phase region and has a liquid phase (oil) of composition a and a vapor phase (gas) of composition a'. As rich gas (LPG) is first injected, it will tend to mix with the liquid a and the overall composition of this mixture could be point α. This mixture α separates into two phases represented by b and b'. As more LPG is injected, it displaces gas b' and mixes with liquid b. These may mix to an overall composition β which separates into liquid c and vapor c'. Injection of more rich gas will result in displacement of the vapor c' and mixing of the liquid c to form the mix δ. This continues until the liquid composition corresponds to that of the critical mixture c^*, at which time a completely miscible displacement is achieved and a single phase would exist in the reservoir.

From the analogy of the vaporizing drive process, it may be easily shown that the leanest injection gas which will give a miscible displacement is represented by a point on the extension of the limiting tie line Ac^* which is tangent at the critical point c^*.

It should now be apparent that an enriched gas process can tolerate low gravity ($\gtrsim 20°$ API) oils, saturated or undersaturated. Also, the injection pressure may be low ($\gtrsim 1000$ psia); however, the lower the pressure, the richer the gas must be. We have also seen that in the event that complete miscibility is not achieved, the concentration gradient of the intermediates allows condensation or transfer of these fractions to the reservoir oil. The latter swells with the added gas; oil viscosity decreases and oil permeability increases, both favoring a more efficient displacement of the oil. Also, gas condensation at the displacement front tends to retard invasion of the oil-saturated portion of the reservoir by the displacing gas, because of the swelling effect at that point and also dissolves the leading fingers of the gas phase. Thus, favorable phase effects may compensate unfavorable viscous forces.

One of the complexities of the condensing gas drive is the readily appreciated fact that it would not be economically feasible to displace the reservoir oil and replace it with an expensive solvent such as LPG. The process has to be modified for use in field operations so that only a small volume of solvent would be required to obtain essentially complete oil recovery. Thus, a slug of solvent is injected which in turn is displaced by a lean gas[13]. Since LPG and methane are highly soluble in each other, miscibility is achieved at lower pressures (critical pressure of LPG-dry natural gas is about 1350 psia at 100° F). However, the main factor to be determined is the minimum size of the LPG slug required to miscibly displace the reservoir oil. This is discussed more appropriately in section 4.5.

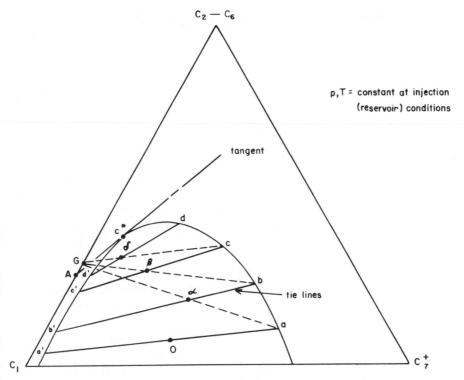

Figure 4.5. Ternary diagram of the reservoir-fluid/displacing-fluid system.

4.3 CO$_2$ Injection

A technique involving the injection of carbon dioxide has received considerable interest for increasing oil recoveries. The mechanism of the carbon dioxide process is essentially similar to that of the vaporizing drive, previously discussed. Carbon dioxide is completely miscible with most reservoir oils only at high pressures just as occurs with dry natural gases. However, carbon dioxide has the added advantage that at moderate reservoir pressures, it is highly soluble (almost of the same order as ethane) in reservoir oils. This causes appreciable swelling and reduction in the viscosity of the oils with the attendant benefits on recovery efficiency.

Moreover, mass transfer between the carbon dioxide and reservoir oil causes extraction of the light hydrocarbons by the CO$_2$-rich phase. Since this extraction takes place above the critical pressure and temperature of CO$_2$ (1070 psia at 88°F), the process is sometimes referred to as retrogade vaporization of the lighter hydrocarbons. The presence of these rich hydrocarbon vapors forming a solvent buffer zone between the CO$_2$ and oil bank, then further enhances oil recovery.

Although there is no economic reason to recover the injected carbon dioxide, nevertheless, because of lack of availability, cost of the liquid* and corrosion problems associated with CO_2, it is desirable that the carbon dioxide injection be a slug process rather than continous. In practice, a slug of CO_2 is injected and followed by a plain or carbonated waterflood. Water is used as the displacing phase because of the rather favorable solubility characteristics of CO_2 in water (6 percent by weight at 957 psia and 79° F).

It has been explained previously that the LPG slug in the condensing gas drive process is usually followed by dry gas injection. In such a process, miscibility must be maintained at both the leading and trailing edges of the LPG slug. Evidently, miscibility conditions at the trailing edge are more critical in that the methane-LPG system may require pressures of the order of 1350 psia. Since this is still too high for shallow reservoirs another form of the basic LPG slug process involves driving the slug with water. Mostly because the LPG-water displacement process is immiscible, however, large amounts of LPG remain unrecovered. More recently, laboratory experiments have indicated the feasibility of following the LPG slug with a water driven CO_2 slugs[36]. Since carbon dioxide is completely or at least partially miscible with propane and soluble in water, thus displacement efficiency is improved at both ends of the CO_2 slug and more LPG is recovered.

The carbon dioxide process has also been approached from still another point of view[16, 23]: using high pressure CO_2 injection to vaporize the reservoir oil. Such a vaporization of reservoir oil by carbon dioxide repressuring appears to have possible use in carbonate reservoirs where the major oil content of the reservoir is contained in the non-fractured porosity with little permeability. The carbon dioxide flows into the fractures, contacts the oil in the matrix and vaporizes part of it. This vaporized oil is produced and recovered and the carbon dioxide is reinjected again.

The carbon dioxide process offers yet another advantage in carbonate reservoirs. The carbonic acid formed by contact with the connate water, dissolves the porous medium hereby increasing permeability. On the negative side, however, this weak acid is the major source of corrosion problems in the well equipment.

4.4 The Dispersion Model

The flow of miscible fluids through porous media is materially different from that of immiscible fluids which we have treated in detail in the previous chapters. Immiscible phases are separated macroscopically by a sharp interface, while miscible fluids have no distinct surface; instead, a transition zone exists between the two fluids (see Figure 4.6).

---

*CO_2 is normally recovered from ammonia synthesis gas or from steam-reforming natural gas at an approximate cost of $4.50 per ton or $0.45 per barrel.

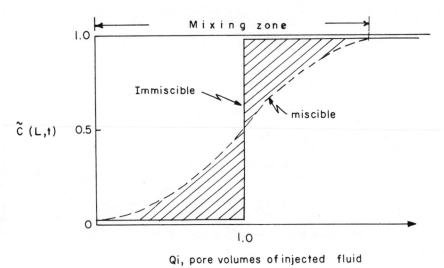

Figure 4.6. Schematic of miscible and immiscible processes.

Thus, as one miscible fluid displaces another, the displacing solution continuously mixes with the resident fluid so that the arrival of the displacing solution at a given point in the porous medium is characterized by a gradual change in the solution concentration from that of the original fluid to that of the invading one. This mixing or interfusing of the two fluids, due to both molecular diffusion and convection, is termed dispersion.

Although Darcy's law describes the macroscopic flow regime within a porous medium, it includes no information on the dispersion phenomena associated with the flow. Several models have been proposed to explain the phenomenon in question; an excellent comprehensive treatment of these is given by Bear[2].

It is generally recognized that the Fick's law model adequately characterizes the mixing of miscible fluids flowing in porous media. Fick's law of molecular diffusion states that the transfer rate of a molecular species is proportional to the concentration gradient. Combining Fick's law with a continuity equation, gives

$$\text{div}\,(D_o\;\text{grad}\;C) = \partial C/\partial t \qquad\qquad 1$$

where D_o is the molecular diffusion coefficient in a given solvent for the species whose concentration is C. Note the similarity of **Equation 1** to the corresponding "diffusivity" equations in unsteady state flow of heat, electricity, and a compressible fluid in a porous medium.

It is of interest to solve **Equation 1** for the semi-infinite linear case when one face is kept at a constant concentration C_o, the initial concentration being zero throughout the medium. If D_o is constant, **Equation 1** becomes

$$D_o \, \frac{\partial^2 C}{\partial x^2} = \frac{\partial C}{\partial t} \qquad\qquad 2$$

The solution is the same as the familiar case of a slab initially at a uniform temperature, with one face suddenly held constant at a different temperature. It is

$$\tilde{C}(x, t) = erfc\left(\frac{x}{\sqrt{4 D_o t}}\right), \quad \tilde{C} = \frac{C}{C_o} \qquad\qquad 3$$

where *erfc* is the complementary error function defined as

$$erfc(y) = \frac{2}{\sqrt{\pi}} \int_y^\infty e^{-w^2} \, dw, \quad y = \frac{x}{\sqrt{4 D_o t}} \qquad\qquad 4$$

Numerical values of this integral are extensively tabulated. Also, a plot of **Equation 3** on arithmetic probability co-ordinate paper yields a straight line. D_o is used to denote molecular diffusion, aside from a porous medium. The solution of **Equation 2** is also readily obtained by the Boltzmann transform method, setting $y = x / \sqrt{4 D_o t}$ and noting the definition of erfc above.

If the miscible fluids are in a porous medium, there are other considerations than the Fick's diffusion constant D_o. If there is general hydrodynamic movement, there is a mixing due to the random viscous motion in the capillary network. This mixing is a combination of diffusion and convection. Then, at any point, there is also a change in composition with time due to the general or overall movement of the mixture past that point, since there is an overall or "mean" concentration gradient. The mixing or "dispersion" constant will be denoted by D_x for linear flow in the x-direction. It depends on the porous medium, properties of the fluids, and the overall velocity of fluid movement as described later.

For linear miscible displacement, then, $\partial C/\partial t$ at a point is governed by two terms, one due to the mixing or dispersion, the other due to the movement of a concentration gradient past that point:

$$D_x \, \frac{\partial^2 C}{\partial_x{}^2} = \frac{\partial C}{\partial t} + U_x \, \frac{\partial C}{\partial x} \qquad\qquad 5$$

$$\underset{\text{axial}}{\underset{\text{diffusion}}{}} \qquad\qquad \underset{\text{axial}}{\underset{\text{convection}}{}}$$

This solution, subject to the previous initial and boundary conditions, has been shown[24] to be

$$\tilde{C}(x,t) = \frac{1}{2}\left\{ erfc\left(\frac{x=\bar{u}t}{\sqrt{4D_x t}}\right) + \exp\left(\frac{\bar{u}x}{D_x}\right) erfc\left(\frac{x+\bar{u}t}{\sqrt{4D_x t}}\right) \right\} \quad 6$$

The second term on the right hand side of **Equation 6** is negligible at some distance from the inflow boundary and therefore we may write

$$\tilde{C}(x,t) \simeq \frac{1}{2} erfc\left(\frac{x-\bar{u}t}{\sqrt{4D_x t}}\right) \quad 7$$

When boundaries are symmetrical about $x = \bar{u}t$ (e.g., in an infinite linear system), the boundary condition $C(-\infty, t) = C_o$ replaces the condition $C(o, t) = C_o$ and **Equation 7** is the exact solution of the symmetrical convection problem. $\bar{u}$ is the average pore or interstitial velocity assumed constant throughout the length of the flow field; $\bar{u}t$ is the average position of the tracer particle, and $x-\bar{u}t$ is then the moving space co-ordinate for the system. Figure 4.7 shows typical concentration profiles during linear miscible displacement.

Equation 7, could also be derived in the manner of **Equation 3** by assuming that the coordinate system moves with velocity u_x, then the last term of **Equation 5** drops out, and, in the solution, real x is $u_x t$ less than the moving x', and $C = \frac{1}{2} C_o$ at $x = \bar{u}t$.

Since some early work on miscible displacement was done in long cylindrical capillaries, it is of interest to examine the mixing mechanism in such a system.

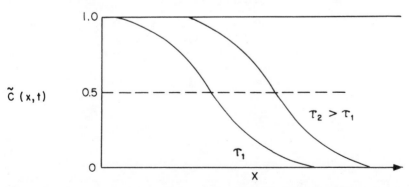

Figure 4.7. Typical concentration profiles during linear miscible displacement.

The general equation is, for a constant D_o

$$D_o \left\{ \frac{\partial C^2}{\partial r^2} + \frac{1}{r}\frac{\partial C}{\partial r} + \frac{\partial^2 C}{\partial_x{}^2} \right\} = \frac{\partial C}{\partial t} + u_r\frac{\partial C}{\partial r} + u_x\frac{\partial C}{\partial x} \qquad 8$$

| radial | axial | radial | axial |
| diffusion | diffusion | convection | convection |

Note that the left side is the Fick (D_o div grad C) term and the last two terms on the right side are due to hydrodynamic movement of a concentration gradient past a point. This equation is evidently very complex for analytic or numerical solution.

If we neglect diffusion in the x (flow)-direction, because of low gradients, and note that, in viscous capillary flow, $u_r = o$, and $u_x = 2\bar{u}\ \left\{ 1 - (r/r_o)^2 \right\}$, **Equation 8** reduces to

$$D_o \left\{ \frac{\partial^2 C}{\partial r^2} + \frac{1}{r}\frac{\partial C}{\partial r} \right\} = \frac{\partial C}{\partial t} + 2\bar{u}\left\{ 1 - \left(\frac{r}{r_o}\right)^2 \right\}\frac{\partial C}{\partial x} \qquad 9$$

where $\bar{u}$ is the average Hagen-Poiseuille velocity in the capillary of radius r_o. Also, the expression for u_x assumes that the viscosities of the displaced and displacing fluids and their mixtures are the same. **Equation 9** is called the Taylor diffusion model and its solution, subject to the same conditions of the infinite linear reservoir, is[35]

$$\bar{C}(x, t) = \frac{1}{2}\ erfc\left(\sqrt{\frac{x - \bar{u}t}{\dfrac{r_o{}^2\ \bar{u}^2 t}{12\ D_o}}} \right),\ x + \bar{u}t > 0 \qquad 10$$

$\bar{C}(x, t)$ is the mean concentration over a cross section, $\bar{C} \sim \tilde{C}$.

Because of the similarity of the solutions, **Equation 7** for pure axial mixing and **Equation 10** for pure radial diffusion effects, this suggests a heuristic solution to include linear flow in a capillary and uniform linear flow in a porous medium, of the form (see p. 000 for another approach).

$$\tilde{C}(x, t) = \frac{1}{2}\ erfc\left(\frac{x - \bar{u}t}{\sqrt{4 D_e t}} \right) \qquad 11$$

where D_e is the effective dispersion coefficient defined as[25]

$$\frac{D_e}{D_o} = \frac{1}{\tau} + a_o P_e{}^m \qquad 12$$

$$P_e = \frac{\bar{u}d_p \epsilon}{D_o}$$ is the Peclet number for Taylor diffusion and measures the degree of importance of convection; a_o is an arbitrary constant depending on pore size distribution; τ is the effective rock tortuosity at the residual water saturation; d_p is the mean grain diameter; ϵ is an inhomogeneity or packing factor, and m indicates the power dependence of dispersion upon velocity. Note that for the capillary system τ is unity and $m = 2$.

Equations 11 and 12 have been experimentally verified[21]. (Figure 4.8) for porous media and m has been found to range from 1.0 to 1.4, with an average value of 1.2. This variation would appear reasonable since it is obvious that the fluid velocity within a porous medium varies widely from the assumed constant average and in places is almost zero.

With regard to the characteristic length $d_p \epsilon$, it varies from 0.23 to 0.75 cm. for sandstones, with an average value of 0.42 cm. Dolomites show higher values. a_o is usually assumed constant with a value of $a_o = 0.5$ and $1/\tau$ varies between 0.15 and 0.7 depending on the lithology of the porous medium.

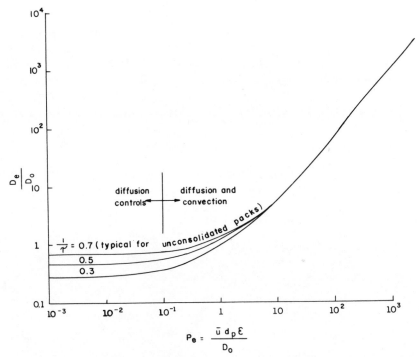

Figure 4.8. Longitudinal dispersion coefficients for porous media (after Perkins et.al.[25]).

Thus **Equation 12**, in its currently accepted form, is

$$\frac{D_e}{D_o} = \frac{1}{\tau} + 0.5 P_e^{1.2}$$

13

which also emphasizes the fact that the apparent molecular diffusion coefficient is less than the value measured in the absence of a porous medium; $D_e = D_o/\tau$ for $P_e = 0$, that is, $\bar{u} = 0$.

Because reservoir-well systems are normally characterized by combinations of linear and radial geometries, it is of interest to examine the dispersion problem in a radial system, in addition to the linear solutions presented thus far.

Hoopes and Harleman[14] have derived the following overall dispersion model for symmetrical radial flow.*

$$\frac{\partial C}{\partial t} + u_r \frac{\partial C}{\partial r} = D'u_r \frac{\partial^2 C}{\partial r^2} + \frac{D_o'}{r} \frac{\partial}{\partial r} \left(r \frac{\partial C}{\partial r} \right)$$

14

$$\underset{\text{convection}}{} \quad \underset{\text{dispersion}}{} \quad \underset{\text{diffusion}}{}$$

Raimondi, et al.[28] suggest that at some distance from the injection well, the influences of dispersion and diffusion on the concentration distribution, as the solvent moves by a point, are small in comparison to the total dispersion and diffusion that has taken place up to that point. Thus, neglecting the right hand side of **Equation 14**, we have

$$\frac{\partial}{\partial t} = - u_r \frac{\partial}{\partial r}$$

15

---

*The corresponding model for linear flow is

$$\frac{\partial C}{\partial t} + \bar{u} \frac{\partial C}{\partial x} = D'\bar{u} \frac{\partial^2 C}{\partial x^2} + D_o' \frac{\partial^2 C}{\partial x^2} = D_e \frac{\partial^2 C}{\partial x^2}$$

a

where

$$D_e = D' \bar{u} + D_o'$$

b

Substituting $\dfrac{\partial}{\partial t} = - \bar{u} \dfrac{\partial}{\partial x}$ in the right hand side of **Equation a**, gives

$$\frac{\partial C}{\partial t} + \bar{u} \frac{\partial C}{\partial x} = D_e \frac{\partial^2 C}{\partial x^2}$$

c

the solution of which is **Equation 11**.

Introducing **Equation 15** into the dispersion and diffusion terms of **Equation 14**, leads to the expression

$$\frac{\partial C}{\partial t} + \frac{q^*}{r} \frac{\partial C}{\partial r} = \left\{ D^1 \frac{r}{q^*} + D_o^1 \frac{r^2}{q^{*2}} \right\} \frac{\partial^2 C}{\partial t^2} \qquad 16$$

where

$$u_r = \frac{q^*}{r}, \quad q^* = \frac{q_w}{2\pi\phi h} \qquad 17$$

For a constant injection rate q_w of a solvent with a concentration C_o at $\gamma = 0$, the solution of **Equation 16** is given by

$$\tilde{C}(r, t) = \frac{1}{2}\, erfc \left\{ \frac{\frac{1}{2} r^2 - q^* t}{f(r)} \right\} \qquad 18$$

where

$$f^2(r) = \frac{4}{3} D^1 r^3 + D_o^1 \frac{r^4}{q^*} \qquad 19$$

D' is termed the intrinsic dispersivity coefficient and is a function only of the rock matrix. Note that D' has the units of length and from analogy with **Equation 12**, it is equivalent to $a_o d_p \epsilon$. Also, D'_o is the apparent molecular diffusion coefficient, that is, $D'_o = D_o/\tau$.

The applications of these linear and radial solutions are discussed in the following section.

Finally, because of the importance of the molecular diffusion coefficient D_o, let us briefly look at the factors which determine its behavior. In particular, we at least would like to qualitatively understand the effects of such parameters as pressure and temperature on the value of D_o.

For a given binary gas system the relationship obtained from kinetic theory is

$$D_o \sim \frac{T^2}{p} \qquad 20$$

which shows that high pressures tend to slow down the diffusion process while high temperatures favor diffusion. In the case of binary liquid systems, we have already pointed out that diffusion there is much slower than in gases.

The following approximate relationship for a given binary system is generally accepted.

$$D_o \sim \frac{T}{\tilde{\mu}} \qquad\qquad 21$$

where $\tilde{\mu}$ is the viscosity of the solution and its presence in the relationship emphasizes the strong dependence of liquid diffusion coefficients on concentration. Some typical values of diffusion coefficients of interest are given in the following table:

system	temp. range, $^\circ$F	pressure range, psia.	D_o – range x 10^5, cm^2/sec.
Methane-Propane (liq)	40-160	367-970	18.2 - 62.5
Methane-Butane (liq)	10-218	265-1250	12.7 - 28.7
Methane-Decane (liq)	40-280	338-4410	1.06 - 18.2
Methane-Crude Oil (liq)	40-280	14.7-4110	1.14 - 11.2
LPG - Crude Oil	at typical reservoir conditions		34
CO_2 - Water	80	14.7	24500

Table 6. Typical Values of Diffusion Coefficients of Interest

It is important to note that although several theoretical and semiempirical relationships for estimating diffusion coefficients in binary gas and liquid systems are available in the literature, these yield rather unreliable results for typical reservoir systems. This is due to the multicomponent nature of petroleum hydrocarbon systems, the high pressures and temperatures involved. All told, diffusion coefficients for reservoir systems must be experimentally determined. Unfortunately, however, experimental difficulties complicate obtaining reliable data especially for liquid systems. A detailed discussion of advances in theoretical and experimental methods for determining diffusion coefficients is presented by Reid and Sherwood[29].

Example 8

It is in the interest of our understanding of the mass transfer process to study the rate at which a volatile oil A evaporates into a vapor B. Assume that the entire system is at rest and at constant pressure and temperature. Such a system may be analogous to the behavior at the gas-oil contact in a petroleum reservoir.

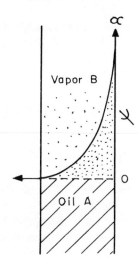

Figure 4.9 Diffusion in a stationary system with a long column of vapor above it.

Solution: For an essentially stationary system, the diffusion **Equation 2**, section 4.4 applies. If we further assume that the column of vapor above the liquid is very long (see Figure 4.9), then **Equation 3**, section 4.4

$$\frac{C_A\ (x,\ t)}{C_{A_o}} = erfc\left(\frac{x}{\sqrt{4\ D_o t}} \right) \qquad\qquad 22$$

represents the solution sought.

We must now calculate the volume rate of production of vapor of A from a surface of area S. If V_A is the volume of A produced by a time t after evaporation has started, then

$$\frac{d\ V_A}{dt} = \frac{N_{A_o}\ S}{C^+} \qquad\qquad 23$$

where N_{A_o} is the molar flux (gm-moles/cm²-sec) of A at $x = O$ and $C^+ = C_A + C_B$ is the molar density of the solution.

Fick's law of diffusion states that

$$N_{A_o} = -D_o \left. \frac{\partial\ C_A}{\partial x} \right|_{x\ =\ 0} \qquad\qquad 24$$

Differentiating **Equation 22**, taking the value of the derivative at $x=0$, and substituting **Equation 24** into **Equation 23** gives

$$\frac{d V_A}{dt} = n_{A_o} S \sqrt{\frac{D_o}{\pi t}}$$

25

which, upon integration, becomes

$$V_A = n_{A_o} S \sqrt{\frac{4 D_o t}{\pi}}$$

26

n_{A_o} is the mole fraction of A at $x=0$ ($n_A = C_A/C^+$).

It is important that we realize that **Equation 26** is only valid for slightly volatile oils since convective transport was ignored. Moreover, the vapors A and B are assumed to form an ideal mixture, hence the molar density C^+ is constant throughout the gas phase.

Consider the oil is contained in a capillary (pore) column of radius 0.01 cm. Further assume $n_{A_o} = 0.5$ and $D_o = 2x\ 10^{-5}$ cm^2/sec.

After one day of mass transfer, the volume of oil evaporated as a vapor at ambient conditions, is

$$V_A = (0.01)^2 \left\{ \pi \times 2 \times 10^{-5} \times 24 \times 3600 \right\}^{\frac{1}{2}} = 2.33 \times 10^{-4} \text{ cc}$$

The calculation shows the slowness of quiescent evaporation by diffusion alone into a large space previously filled with a gas. The establishment of a gas cap in equilibrium with an oil zone over geologic time may be related to such a mechanism. However, in the operation of a field, we are dealing with gas and oil together in microscopic pores and, while diffusion is part of the mechanism of establishing any equilibrium, any departure from equilibrium due to diffusion rates is so insignificant that practical calculations are based on phase equilibria (e.g., flash or differential). The large amounts of gas often required to vaporize oils (as much as 5 pore volumes) are readily accounted for on the basis of phase relations. In depletion drives, allowance is sometimes made for a "lag factor," but this is due to a tendency of the liquid phase to be supersaturated, that is, a lag in releasing the gas to give equilibrium conditions.

4.5 Optimizing the Solvent Slug Size

In the condensing gas drive process, oil is usually displaced by a small amount, or slug, of solvent. Dry gas in turn follows the solvent slug. The design of such a miscible flood is critical in three aspects.

In the first place, the engineer must determine the maximum permissible dilution of the solvent for miscibility since economics dictate that we should blend the slug as much as possible. This information is available from the ternary diagram and has been already discussed in section 4.2.

Secondly, the size of the solvent slug must be an optimum in the sense that it cannot be too large, again for economic reasons, nor too small because of the physical requirement of maintaining miscibility in spite of degradation by mixing as it moves through the reservoir. Finally, control of the composition of the injection solvent can become a serious and difficult problem, particularly in cases where the enrichment material is being furnished from a number of field producing gas streams.

This section is principally concerned with the second problem, that is, the determination of the optimum slug size; the third problem is briefly discussed at the end of the section.

From a practical point of view, optimizing the solvent slug size in fact implies determining the minimum size slug that can be injected for efficient displacement of the reservoir oil. This minimum is essentially determined by those factors which cause deterioration of the solvent bank as it moves through the reservoir from injection to producing well. When these factors cause the concentration within the slug to drop below the level required to maintain misciblity, a less efficient immiscible displacement prevails during the remainder of the flood.

There appear to be four principal factors which bring about deterioration of the solvent slug: dispersion, channeling, viscosity and density differences of the injected and resident fluids. These factors operate on different scales and it is important that we make a distinction.

On the localized scale, the solvent slug mixes or blends with the fluids immediately preceding and following it in accordance with the dispersion process. Transverse mixing causes loss of solvent to by-passed zones while longitudinal mixing stretches out the transition or mixing zone thereby decreasing its piston-like effectivity.

On the overall macroscopic scale, involving the general hydrodynamics of the flow process, the solvent may by-pass the oil flowing through permeable streaks, by overriding the oil because of density difference, or by fingering because of unfavorable mobility (viscosity) ratios. In such cases, the increase in effective area available for dispersion of the rich components leads to a faster dissipation of the solvent slug.

In general, then, these four factors promote a large mixing zone which is detrimental to the miscible process.

Let us first examine the case of miscible drive in an idealized porous medium consisting of homogeneous fluids, that is, fluids of equal viscosity and density.

For a system of linear geometry, **Equation 11**, section 4.4 applies. It describes the growth of the mixing zone, that is, the zone in which $\tilde{C}$ varies from essentially zero to essentially unity (see Figure 4.7). Theoretically, because of the asymptotic behavior of the complementary error function, the distance from 0 to 1 is infinite, so it is impractical to speak of this as the length of the zone. Since the S-shaped curve becomes quite flat as $\tilde{C}$

approaches zero or unity, it is customary to select arbitrary values of $\tilde{C}$ for the zone boundaries, such as 0.05 and 0.95 or 0.10 and 0.90. Note that the curve is symmetrical about $x = \bar{u}t$ and that $x - \bar{u}t$ measures the distance (positive or negative) of $\tilde{C}$ from the center of the zone.

Noting that the numbers whose erfc's are 0.2 and 1.8 are, respectively, 0.9062 and -0.9062, **Equation 11**, section 4.4 gives

$$\Delta \tilde{x} = x_{.1} - x_{.9} = 1.8124 \sqrt{4 D_e t} = 3.625 \sqrt{D_e t} \qquad 1$$

where $\Delta \tilde{x}$ is the length of the transition zone and $x_{.1}$ and $x_{.9}$ are the positions of the 10 and 90 percent concentrations, respectively.

For a certain system and constant velocity, then, the length of the transition zone increases as the square root of the distance traveled by the middle of the zone, $x_{.5}/\bar{u}$. Except at extremely low velocities, in the case of liquids the diffusion term in **Equation 13**, section 4.4, is small compared to the P_e term. Substituting for P_e, **Equation 13** section 4.4, may be written, for a given system, $D_e \sim \bar{u}^{1.2}$. Substituting this and $x_{.5}/\bar{u}$ for t in **Equation 1**, obtains

$$\Delta \tilde{x} \sim x_{.5}^{\frac{1}{2}} \bar{u}^{0.1} \qquad 2$$

Thus, for a given distance traveled by the zone center, the length of the mixing zone in most cases is only slightly increased by flowing at a higher velocity. It has also been found that the growth of the transition zone is not greatly affected by the viscosity if displacing and displaced liquids have nearly the same viscosity. However, the growth is reduced by displacing a liquid by a liquid of higher viscosity and increased by displacing with a liquid of lower viscosity. In the latter case, however, "viscous fingering" may take place, as will be discussed later.

It should be emphasized that **Equation 11**, section 4.4, is strictly for the case of single phase miscible displacement without heterogeneities or viscous fingering. In the case of an enriched gas drive in a petroleum reservoir, diffusion is involved as in all interphase transfer processes, but the macroscopic multiple contact-equilibrium factors create the "front" and will completely dominate the mechanism. In the case of a gas injected into a gas reservoir, **Equation 11**, section 4.4, would be valid if complete homogeneity existed.

If, for instance, we take $\tau = 1.6$ $\bar{u} = 1$ ft/day (3.5×10^{-4} cm/sec), and $D_o = 0.1$ cm^2/sec, the molecular diffusion term in **Equation 13**, section 4.4, completely predominates. However, because of the low viscosity of gases, slight variations, at field dimensions, in density, temperature, effective permeabilities, etc., will cause flow distributions that will far overshadow molecular diffusion effects.

Up to this point we have been studying the solvent slug dissipation only at its leading edge. In effect, we have assumed that solvent injection is continuous which is sometimes referred to as injection of an infinite slug. For a finite slug, its trailing edge is also degraded by dispersion of the pusher gas. More correctly then, **Equation 11**, section 4.4, should be modified to account for injection of a finite rectangular slug.

By the superposition theroem, the modified solution of **Equation 11**, section 4.4, can be easily shown to be

$$\tilde{C}\,(x,\,t) = \tfrac{1}{2}\left[\,erf\left\{\frac{x-\bar{u}\,(t-t_1)}{\sqrt{4\,D_e\,(t-t_1)}}\right\} - erf\left\{\sqrt{\frac{x-\bar{u}\,t}{4\,D_e\,t}}\right\}\,\right]\qquad 3$$

where erf is the error function, $erf = 1 - erfc$. t_1 is the time for which pure solvent is assumed to have been injected (see Figure 4.10).

If we measure distances always from the front of the slug, that is, assuming a moving coordinate system so that $x' = x - \bar{u}t$, **Equation 3**, can be plotted as shown in Figure 4.11. This plot shows the loss of solvent concentration as the slug moves through the formation. Thus, we can establish a criterion for

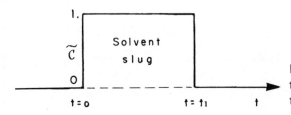

Figure 4.10. Degradation of the trailing edge of a finite slug by the dispersion of the pusher gas.

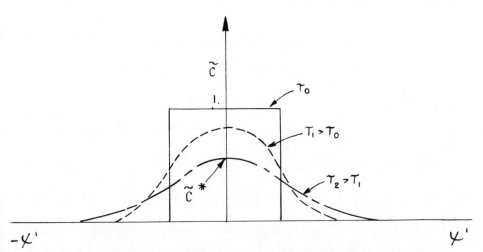

Figure 4.11. Loss of concentration as a finite slug moves through a formation.

determining the optimum slug size, namely, that the maximum solvent concentration reached at any distance must be at least equal to the minimum solvent concentration required for miscibility, as obtained from the ternary composition diagram.

It is assumed that the maximum solvent concentration occurs at a time half way between the times at which the head and tail of the slug pass a given point. From **Equation 3**, we therefore get

$$\widetilde{C}^* (x) = erf \left(\frac{\widetilde{\Delta x_i}}{\sqrt{16 \, x \, D_e/\bar{u}}} \right) \qquad\qquad 4$$

where $\widetilde{\Delta x_i}$ is the initial length of the pure solvent slug. If we interpret C^* as the critical concentration required for miscibility, then **Equation 4** defines the distance x which a slug of initial size $\widetilde{\Delta x_i}$ would travel before miscible breakdown.

For the equivalent problem in radial geometry, **Equation 18**, section 4.4, modifies to

$$\widetilde{C} (r, t) = \tfrac{1}{2} \left[erf \left\{ \frac{\tfrac{1}{2} r^2 - q^* (t - t_1)}{f (r)} \right\} - erf \left\{ \frac{\tfrac{1}{2} r^2 - q^* t}{f (r)} \right\} \right] \qquad 5$$

when considering the injection of a finite slug. By the same token, the maximum solvent concentration can be shown to be

$$\widetilde{C}^* (r) = erf \left(\frac{\widetilde{\Delta r_i}^2}{\sqrt{\dfrac{64}{3} D^1 r^3}} \right) \qquad\qquad 6$$

where $\widetilde{\Delta r_i}$ (from $r = 0$) is the length (radius) of the undiluted slug just after injection.

Note that from **Equation 6** the slug requirement, expressed as the ratio of slug volume to pore volume swept, is inversely proportional to the square root of the pore volume swept for a fixed concentration $\widetilde{C}^*$, that is, $\dfrac{\widetilde{\Delta r_i}^2}{r^2} \sim r^{-\frac{1}{2}}$ For the linear system, **Equation 4** indicates a similar relationship, namely, $\dfrac{\widetilde{\Delta x_i}}{x}$ $\sim x^{-\frac{1}{2}}$. Thus, from the models presented, it appears that slug deterioration is independent of geometry. This, although not physically true, would seem to indicate that the linear model is a fair representation of the reservoir system.

Our discussion to this point has been concerned with displacement in a homogeneous reservoir containing homogeneous fluids. As is shown in Example 9, under these conditions a solvent slug of the order of 2 to 3 percent of the hydrocarbon pore volume should be sufficient for complete miscible displacement.

Unfortunately, most field miscible displacement processes involve the displacement of oils by solvents of much lower viscosity and density. The displacement process at these adverse viscosity and density ratios is dominated by instability phenomena such as viscous fingers and gravity tongues. The effects of permeability stratification and heterogeneity are equally adverse. These phenomena dilute the slug and increase slug requirements.

Consider also the radial problem mentioned earlier. In this case transverse dispersion dominates over longitudinal dispersion. The transverse dispersion would now suppress viscous fingering but the reduced longitudinal dispersion sharpens the front, increasing viscosity contrast between the displacing and resident fluids.

We can therefore appreciate that the field problem suddenly becomes extremely complex, to the extent that the entire concept of a mixing zone is no longer valid. The same is true for **Equations 3** and **5**. It is as though we went from laminar to turbulent flow in immiscible hydrodynamic systems.

However, in order to make the problem mathematically tractable, it is necessary that we retain the idea of a mixing zone, assuming a kind of average behavior for the system. As a matter of fact, both experimental and field evidence indicate that in the presence of instabilities, the length of the "mixing zone" behaves like $\widetilde{\Delta x} \sim L^n$ where $n \geq 1$, as compared to $n = \frac{1}{2}$ for the homogeneous problem, as given by **Equation 1**. This would also imply the need for large solvent slugs, usually of order of at least 10 to 15 percent hydrocarbon pore volume.

One of the most comprehensive treatments of the unstable problem was presented by Perrine[26]. He used perturbation methods to define the conditions under which the displacement may be stabilized through either rate control so that gravity may dominate the viscous forces, or by using a graded viscosity zone to stabilize the flow at some pre-determined injection rate. The method of rate control is not usually practical because of the low flow rates required and will not be discussed further.

The gradient technique uses a gradual change in the properties of the injected fluid rather than an abrupt change from one fluid to another, in order to attenuate the viscous or density differences that cause instabilities. The derivation of Perrine's model is beyond the scope of this monograph and therefore we will limit ourselves to outline the relevant results, which are valid for thin sands.

The condition for neutral stability or in other words the condition under which instabilities do not grow within the system, also defines the minimum size of the solvent slug for stable displacement. This condition is specified as

$$\frac{\partial \widetilde{C}^*}{\partial x} = \frac{\widetilde{n}^2 \, \pi^2 \, De}{\bar{u} \, h^2 \, g_1}, \quad \widetilde{n} = 1, 2, 3, \ldots \qquad 7$$

where

$$g_1 = \frac{d \ln \widetilde{\mu}}{d \widetilde{C}} + \frac{k}{\widetilde{\mu} \, \bar{u}} \, \frac{d \widetilde{\rho}}{d \widetilde{C}} \, g \, \text{Sin} \, \alpha \qquad 8$$

$\widetilde{n}$ is a perturbation property, characteristic of vertical permeability variation and may be approximated by Koval's heterogeneity factor, $\widetilde{n} \sim \widetilde{V}$, (section 4.6); a conservative estimate of $\widetilde{n}$ is unity. The asterisk on the concentration variable indicates a critical, or limiting, relationship as perhaps defined by the phase diagram. h is the formation thickness and the function $g_1 \, (\widetilde{C})$ measures the relative dominance of gravity over the viscous forces. Viscosity $\widetilde{\mu}$ is for a fluid mixture of solvent concentration $\widetilde{C}$; k is permeability and α is the formation dip, negative for downward tilt in the direction of flow.

Equation 7 simply defines the steepest solvent concentration gradient that must be injected in order to maintain stability. The time t^* required to inject this concentration gradient is obtained by using the following equivalence

$$\frac{\partial \widetilde{C}^*}{\partial x} = -\frac{1}{\bar{u}} \, \frac{\partial \widetilde{C}^*}{\partial t} \qquad 9$$

to integrate **Equation 7**. The resultant expression is

$$t^* = -\frac{h^2}{\widetilde{n}^2 \, \pi^2 \, D_e} \int_o^{\widetilde{C}^*} g_1 \, d\widetilde{C} \qquad 10$$

for zero initial conditions.

Although this concept of the injection of a solvent gradient is sound, it involves operational difficulties[+] which limit its practical application. A simplified alternate procedure would be to replace the injection of a solvent

---

[+] In the field it would be necessary to mix two injection streams at pumping rates varying uniformly with time to keep the total flow rate as constant as possible, and to give the desired concentration gradient.

gradient by injection of pure solvent. Thus, the injection profile becomes rectangular in shape and contains a solvent excess that depends in an appropriate way on the degree of instability. The problem is therefore reduced to the behavior of a rectangular slug with stable flow (see Figure 4.12) and is handled as follows.

Let us consider the complete problem in which gas displaces a rectangular solvent slug. The pertinent solvent and gas concentration equations, along the lines of **Equation 3**, are

$$\widetilde{C} = \frac{1}{2} \left[erfc \left\{ \frac{x - \bar{u}t}{\sqrt{4 D_e t}} \right\} - erfc \left\{ \frac{x - \bar{u} \ (t - t')}{\sqrt{4 D_e \ (t - t')}} \right\} \right] \qquad 11$$

and

$$\widetilde{C}_g = \frac{1}{2} erfc \left\{ \frac{x - \bar{u} \ (t - t')}{\sqrt{4 D_e \ (t - t')}} \right\} \qquad 12$$

where $\widetilde{C}_g$ is the fractional gas concentration and $1 - \widetilde{C} - \widetilde{C}_g$ is the oil concentration.

The first step in this approximate method is to compute $\widetilde{C}$ and $\widetilde{C}_g$ using **Equations 11** and **12** for various positions x within the transition zone. The time t' is the time for which pure solvent is assumed to have been injected. Trail-and-error procedures are used to find the smallest time t' for which all portions of the solvent slug will remain miscible until it has reached the end of the system.

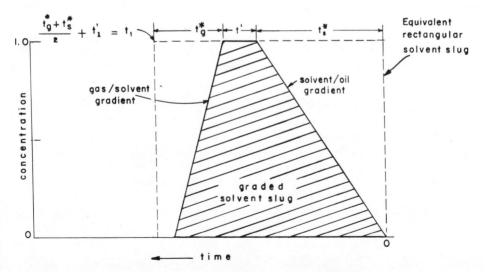

Figure 4.12. Behavior of a rectangular slug with stable flow.

Next, from **Equation 10**, the time, t^*_s, required to inject a graded solvent-oil transition zone is computed. The computation is repeated to find the time, t^*_g, to inject the gas-solvent transition zone. The function g_1 will differ from the two calculations.

Finally, the estimated time during which solvent alone should be injected to form an extended rectangular slug which includes an allowance for instability, is the sum of t' and the average of the two values of t^*, previously calculated.

$$t_1 = t' + \tfrac{1}{2}\,(t^*_s + t^*_g) \hspace{3cm} 13$$

It should be clear that although this approximate method makes a reasonable allowance for instability, the abrupt concentration change may itself give rise to fingering and breakdown of the rectangular slug. Current experimental evidence[20], however, seems to indicate that the estimation technique is sufficiently conservative. As shown in Example 9, the solvent requirements increase from 2 percent of a pore volume for homogeneous considerations to 20 percent when instability phenomena are taken into account.

We will now briefly make reference to other approaches, mostly based on experimentally developed correlations, that provide estimates of the size of the solvent slug to be injected.

In the absence of gravity effects, viscosity difference is undoubtedly one of the important variables influencing the length of the mixing zone. Fitch and Griffith[11] found a very simple empirical equation which essentially adjusts the theoretical square root relationship of the linear homogeneous system, through fluid viscosity level and ratio. Their equation is

$$\frac{\widetilde{\Delta x}}{L} = \left\{ \frac{\Delta \mu_{o-s}}{10 L \mu_s / \mu_o} \right\}^{1/2} \hspace{3cm} 14$$

where the length of the system L must be expressed in feet and viscosity in centipoise. The subscripts s and o refer to solvent and oil, respectively.

Another important factor in deterioration of a small solvent slug is permeability channelling. The effect of permeability stratification is to propel the solvent through the most permeable zones, bypassing the tighter strata. This exposes the comparatively small slug to large areas of fresh resident fluid causing dissipation of the solvent as it mixes with fluids in adjacent zones. Koonce and Blackwell[18] developed a generalized graphical correlation that permits estimation of the slug size in a stratified reservoir containing homogeneous fluids. Their results can be conveniently combined with, say, Fitch and Griffith's method which relates non-homogeneous fluid behavior.

To finalize the discussion of this section we will briefly touch on another important operational aspect of the miscible displacement problem. We refer to composition control of the injection solvent when the latter is supplied from different field streams.

One evident way to effect control of the injection gas composition is through frequent or continuous sampling with an on-line chromatograph. In many cases, however, this procedure may be impractical (as in remote areas) and an alternative would be to run vapor-liquid equilibrium calculations on the different field separator conditions and establish more easily controlled criteria such as the minimum ratio of rich gas to casing head (dry) gas. Because of the not-too-high pressures involved in solvent slug projects, the vapor-liquid equilibrium calculations are reliable[10].

An even simpler method is suggested by Rutherford[31] who found that at constant pressure, miscibility is a function only of the pseudo critical temperature of the injection gas. This significant result implies that it is only necessary to determine the pseudo critical temperatures of the series of gas mixtures to be encountered in the supply streams. The technique, likewise, simplifies appreciably the laboratory determination of the limits of miscibility in sand packs.

Example 9a

Calculate the minimum size slug of LPG required to displace miscibly a medium grade reservoir oil. Consider the reservoir to be horizontal, linear and homogeneous with a distance of 200 feet between injection and production wells. Injection rate is approximately 3.47×10^{-4} cm/sec (~1 ft/day); μ (gas) = 0.02 cp; μ (LPG) = 0.10 cp; μ (oil) = 0.5 cp; D_o (gas-LPG) = 50×10^{-5} cm²/sec; D_o (LPG-oil) = 30×10^{-5} cm²/sec; τ = 1.6; $d_p \epsilon$ = 0.42 cm. The problem is depicted in Figure 4.13.

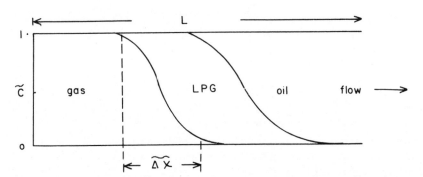

Figure 4.13. Injection of the minimum size slug of LPG required to displace a medium grade reservoir oil in a horizontal, linear and homogenous reservoir.

Solution: For this case we consider the reservoir to contain homogeneous fluids. **Equation 11**, section 4.4, which assumes continuous solvent injection, applies. We first compute the effective dispersion coefficients for the system. From **Equation 13**, section 4.4,

$$D_e \text{ (LPG-oil)} = \left\{ \frac{1}{1.6} + 0.5 \left(\frac{0.42 \times 3.47 \times 10^{-4}}{30 \times 10^{-5}} \right)^{1.2} \right\} 30 \times 10^{-5}$$

$$= 25 \times 10^{-5} \text{ cm}^2/\text{sec}$$

and

$$D_e \text{ (gas-LPG)} = \left\{ \frac{1}{1.6} + 0.5 \left(\frac{0.42 \times 3.47 \times 10^{-4}}{50 \times 10^{-5}} \right)^{1.2} \right\} 50 \times 10^{-5}$$

$$= 37 \times 10^{-5} \text{ cm}^2/\text{sec}$$

From **Equation 1**, the r..tive lengths of the mixing zones are

$$\left. \frac{\widetilde{\Delta x}}{L} \right|_{\text{LPG-oil}} = 3.625 \left(\frac{D_e}{\bar{u}L} \right)^{\frac{1}{2}} = 3.625 \left(\frac{25 \times 10^{-5} \times 24 \times 3600}{1 \times 200 \times (30.48)^2} \right)^{\frac{1}{2}}$$

$$= 0.0391$$

and

$$\left. \frac{\widetilde{\Delta x}}{L} \right|_{\text{gas-LPG}} = 0.0475$$

Therefore the average slug size required is ½ (0.0391 + 0.0475) = 0.043 pore volume.

Note that D_e in the above equation theoretically refers to the transverse dispersion coefficient which is defined by **Equation 13**, section 4.4, with a_o = 0.0157 instead of 0.5. A conservative estimate, however, is simply $D_o/\tau = 30 \times 10^{-5}/1.6$ as was used above.

Example 9b

Rework the previous problem using the correlation of Fitch and Griffith.

Solution: For the solvent-oil system **Equation 14** gives

$$\left.\frac{\widetilde{\Delta x}}{L}\right|_{\text{LPG-oil}} = \left\{\frac{\Delta \mu_{o\text{-}s}}{10 L \mu_s/\mu_o}\right\}^{\frac{1}{2}} = \left\{\frac{0.5 - 0.1}{10 \times 200 \times .1/.5}\right\}^{\frac{1}{2}} = 0.0315$$

and for the gas-solvent system

$$\left.\frac{\widetilde{\Delta x}}{L}\right|_{\text{gas-LPG}} = 0.0141$$

Taking the average of both values, gives a slug size of 0.0228 pore volume.

Example 9c

Repeat the same problem using Perrine's approximate method. Assume a formation thickness of 3 ft. and $\widetilde{n} = 2$. Assume also that phase behavior is known as a function of the solvent concentration as shown in Table 7. Also, **Equations 11** and **12** must be interpreted to be expressed in terms of volume fractions in accordance with the phase data. Viscosity of the three component miscible phase is conveniently given as an empirical equation, namely

$$\widetilde{\mu} = 0.5 \exp. (-1.5 \, \widetilde{C} - 3.2 \, \widetilde{C}_g) \qquad\qquad 15$$

corresponding to the 0.5 cp oil and 0.02 cp gas.

Table 7. Limiting Compositions for Miscibility[26]		
if $\widetilde{C}_g$ is	$\widetilde{C}$ must be $\geqslant$	$(1 - \widetilde{C} - \widetilde{C}_g)$ must be $\leq$
1.00	0	0
0.97	0.0291	0.0009
0.94	0.0582	0.0018
0.92	0.0774	0.0026
0.90	0.0966	0.0034
0.89	0.1062	0.0038
0.88	0.1157	0.0043
0.87	0.1251	0.0049
0.86	0.1344	0.0056
0.85	0.1435	0.0065
0.84	0.1524	0.0076
0.83	0.1611	0.0089
0.82	0.1695	0.0105
0.81	0.1777	0.0123
0.80	0.1854	0.0146

Solution: From **Equations 8** and **10** we compute the time required to inject the graded LPG slug. Thus

$$t_s^* = -\frac{h^2}{\widetilde{n}^2\,\pi^2 D_e} \int_0^1 g_1 d\widetilde{c} = -\frac{h^2}{\widetilde{n}^2\,\pi^2\,D_e} \int_{\widetilde{\mu}=\frac{1}{2}}^{0.112} d\,ln\widetilde{\mu} \qquad 16$$

$$= \frac{(\,3\ x\ 30.48)^2\,(-1.5)}{4\pi^2\ x\ 18.75\ x\ 10^{-5}} = 17\ x\ 10^5\ sec\ \cong\ 20\ days$$

Similarly, $t_g^* = 22$ days for the graded pusher gas-LPG transition zone. Note that in calculating the viscosity from **Equation 15**, $\widetilde{C}_g = 0$ at the leading edge of the slug and $\widetilde{C} = 0$ at the trailing edge.

We must now compute, using **Equations 11** and **12**, the smallest time t' for which all portions of the LPG slug will remain miscible throughout the displacement process. Let t be the time required for the flood to progress through the entire system (i.e. $t \simeq 200$ days). Using a trial value t', compute $\widetilde{C}$ and $\widetilde{C}_g$ for several positions x. If all $\widetilde{C}$ and $\widetilde{C}_g$ values correspond to a single fluid phase as given in **Table 7**, the trial t' overestimated the amount of solvent needed. A smaller t' should then be used. On the other hand, if the data indicates that two fluid phases would be present at some point in the system, a larger value of t' should be used.

For this problem $t' \cong 20$ days and therefore the total time during which LPG alone should be injected to form an extended rectangular slug is

$$t_1 = 20 + \frac{1}{2}\ (20 + 22) = 41\ days$$

or about 20% pore volume.

4.6 Performance Predictions for Miscible Drives

In the determination of the performance of any fluid injection project, miscible or immiscible, we must thoroughly consider the behavior of three efficiency factors. These are displacement, areal and vertical sweep efficiencies. As we discussed in the previous section, miscible flooding in petroleum reservoir systems are inherently unstable because of unfavorable mobility ratio and unfavorably gravity differences between the displacing (solvent) and displaced fluids. Also, channeling due to permeability stratification adversely affects the fluid distribution of the injection fluid. Thus, although in theory the displacement efficiency is 100 percent in miscible displacement processes, in practice viscous fingering and gravity overriding of the injection fluid reduce considerably actual in-place fluid removal. By the same token, fingering and channeling bring on poor areal and vertical sweep efficiencies so that as a whole, the efficiency and economics of miscible drives depend critically on the degree of instability of the system.

Several models for predicting the behavior of miscible floods under unstable conditions have been proposed in the literature. These have taken two approaches: (1) development of a flow model akin to the Buckley-Leverett-Dietz approach for immiscible displacement but combining the effects of flow and mixing, and (2) simultaneous application of Darcy's law for flow and the diffusion-convection equation for mass transfer.

The Buckley-Leverett-Dietz type model developed by Koval[19] has proven to be very reliable while conserving the virtue of simplicity and an intuitive relationship to, perhaps, the more familiar immiscible displacement theory. We will therefore examine Koval's model in this discussion.

Consider a horizontal and linear reservoir system into which a solvent, miscible in all proportions with the resident oil, is continuously injected. For the corresponding immiscible displacement problem the fractional flow formula (see section 3.1) is

$$f_s = \left(1 + \frac{k_o}{k_s} \frac{\mu_s}{\mu_o} \right)^{-1} \qquad\qquad 1$$

and the pore volumes of injected fluid, Q_i, is simply related by

$$Q_i = \left(\frac{df_s}{dS} \right)^{-1} \qquad\qquad 2$$

where the subscript s refers to the injected fluid (solvent in the miscible analogy), and S is its saturation.

Equation 1 can be written in a more general way

$$f_s = F\left(S, \widetilde{V}, E\right) \qquad\qquad 3$$

in which $\widetilde{V}$ stands for heterogeneity effects and E, for vscosity effects. In ordinary immiscible displacement, the relative permeabilty ratio, k_o/k_s, contains both the effects of heterogeneity and of saturation and E is simply the actual viscosity ratio.

In a miscible drive, however, if we postulate that the solvent and oil retain their individual identities, then $k_o = kS_o$ and $k_s = kS_s$, or

$$\frac{k_o}{k_s} = \frac{1 - S}{S} \qquad\qquad 4$$

where S, in effect, is the fraction of the pore volume occupied by the pure solvent and $(1 - S)$ is the fraction occupied by the pure oil (see Figure 4.14).

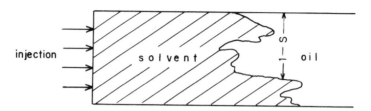

Figure 4.14. Ratio of the fraction of the pore volume occupied by the pure solvent to the fraction occupied by the pure oil in a miscible drive in which the solvent and oil retain their individual identities.

Note that the effects of heterogeneity are no longer associated with **Equation 4.** In regard to the viscosity differences between the solvent and reservoir oil, experimental evidence shows that the viscosity ratio of the pure materials, μ_s/μ_o, is much higher than the effective viscosity ratio during miscible displacement.

From the foregoing discussion and by anology with **Equation 1**, Koval suggests the following fractional flow formula

$$f_s = \left\{ 1 + \left(\frac{1-S}{S} \right) \left(\frac{1}{\widetilde{V}} \right) \left(\frac{\mu_s}{\widetilde{\mu}} \right) \right\}^{-1} = \left\{ \left(1 + \frac{1-S}{S} \right) \left(\frac{1}{\widetilde{V}} \right) \left(\frac{1}{E} \right) \right\}^{-1} \qquad 5$$

to describe the general miscible displacement process. The parameter $\widetilde{V}$ accounts for rock inhomogeneities, dispersion and channeling. It is related to the Dykstra-Parsons permeability variation factor V (see section 3.9), as shown in Figure 4.15. E is the effective viscosity ratio defined as

$$E = \left\{ 0.78 + 0.22 \left(\frac{\mu_o}{\mu_s} \right)^{1/4} \right\}^4 \qquad 6$$

which is just a convenient form of the more general mixing rule

$$\mu = \left\{ \widetilde{C} \mu_s^{-1/4} + (1 - \widetilde{C}) \, \mu_o^{1/4} \right\}^{-4} \qquad 7$$

with $\widetilde{C} = .22$ as suggested by experimental data.

If we define $M = \widetilde{V}E$, **Equation 5** can be written as

$$f_s = \frac{MS}{1 + S(M - 1)} \qquad 8$$

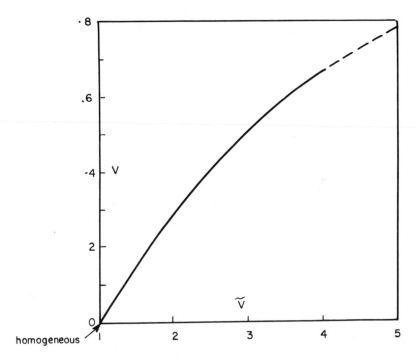

Figure 4.15. Relationship between Koval's(V) and Dykstra-Parson's(V) heterogeneity factors (after Koval[19]).

Differentiating **Equation 8** and combining the result with **Equation 2**, gives

$$f_s = \frac{M - (M/Q_i)^{1/2}}{M - 1} \qquad\qquad 9$$

which is valid at and beyond breakthrough of the injected solvent. Thus, for a given solvent cut, f_s, the cumulative pore volumes of solvent required to be injected is obtained directly from **Equation 9**.

At breakthrough $f_s = 0$ and

$$(Q_i)_{bt} = \frac{1}{M} \qquad\qquad 10$$

Likewise, the pore volumes required to completely displace all of the reservoir oil ($f_s = 1$) is given by

$$(Q_i)_{total} = M \qquad\qquad 11$$

Oil recovery in pore volumes, $\widetilde{N}_p$, can be obtained directly by integrating the f_o versus Q_i curve. Thus

$$\widetilde{N}_p = (Q_i)_{bt} + \int_{(Q_i)_{bt}}^{Q_i} (1 - f_s)\ dQ_i \qquad\qquad 12$$

Introducing **Equation 10** in **Equation 12** and integrating yields

$$\widetilde{N}_p = \frac{2\,(MQ_i)^{\frac12} - 1 - Q_i}{M - 1} \qquad\qquad 13$$

Because of the simplicity of the two fundamental **Equations 9** and **13**, universal curves for predicting reservoir performance may be easily prepared. A plot of **Equation 13** is shown in Figure 4.16.

In a manner analogous to immiscible displacement theory, there we observed that gravitational effects can stabilize an unstable displacement process if the total volumetric flow rate per unit area is less than a critical value. The same is true for miscible floods. Moreover, during the previous discussion of Koval's model, it may have occurred to the reader that the definition of relative permeability, as given by **Equation 4**, is identical to that used in the Dietz model (section 3.6) for immiscible fluid flow.

From a heuristic point of view, then, we may postulate that the fractional flow formula for miscible floods with gravity effects, should follow directly from **Equation 15**, section 3.6. By analogy

$$f_s = \frac{1 - \dfrac{k\,(1-S)}{u\widetilde{\mu}}\ g\,\Delta\widetilde{\rho}\ \dfrac{\text{Sin }\beta}{\cos\,(\alpha - \beta)}}{1 + \dfrac{1 - S}{SM}} \qquad\qquad 14$$

Similarly, the critical velocity, u_c, at which instability occurs, may be deduced from the corresponding equation for immiscible displacement, **Equation 2**, section 3.7. Thus

$$u_c = \left| \frac{kg\Delta\widetilde{\rho}\ \text{Sin }\alpha}{\widetilde{\mu} - \mu_s} \right| \qquad\qquad 15$$

where it is assumed that $k_o = k_s = k$ as occurs during stable flow. Also, $\Delta\widetilde{\rho} = \widetilde{\rho} - \widetilde{\rho}_s$ and

$$\widetilde{\rho} = C\rho_s + (1 - \widetilde{C})\ \rho_o \qquad\qquad 16$$

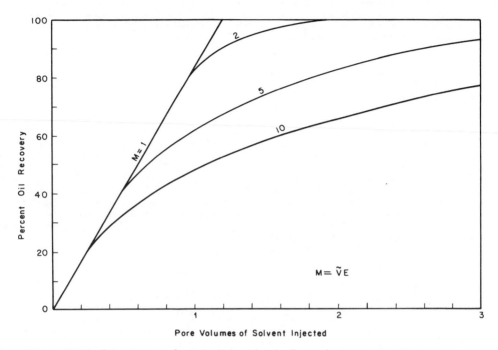

Figure 4.16. Oil recovery for miscible drive in linear heterogeneous reservoirs.

The critical condition indicated by **Equation 15** can also be obtained by setting the function $g_1 = 0$ in **Equation 8**, section 4.5 of Perrine's perturbation model and discretizing the differential terms. Likewise, **Equations 14** and **15** are similar to those derived by Dougherty[8].

Finally, although these models were developed for linear systems, this imposes no serious limitation if we need to evaluate nonlinear systems such as pattern floods. The extension may be easily accomplished by including the appropriate areal coverage factors.

Example 10

A semi-volatile oil reservoir, originally saturated at 3000 psia, was depleted to 1250 psia while producing 20 MMSTB of oil and 22.8 MMMSCF of associated gas. It is estimated, by the material balance methods of Chapter 2, that 10 MMSTB additional oil could be obtained over a 15-year period by straight depletion to economic abandonment for a total net profit of $12 MM ($1.20/barrel).

We desire to analyze the economic feasibility of high pressure gas injection to miscibly displace the remaining oil. To accomplish this, it is proposed to shut-in the reservoir for five years during repressuring to 4500 psia. The reservoir is then produced for eight years at 4500 psia; then injection is stopped and the reservoir is blown-down (depleted) for the next two years.

The saturated fluid properties, at reservoir temperature of 250°F, are

p, psia:	1250	1500	2000	3000	4500
B_o, RB/STB:	1.225	1.250	1.30	1.40	1.55
R_s, SCF/STB:	450	500	600	800	1100

Other data:

Original oil in place, N = 100 MMSTB; μ_o/μ_g = 10;
Permeability variation, V = 0.50; B_g = 2.48/p, RB/SCF

Analysis:

Original gas in place = NR_{si} = 100 x 800 = 80 MMMSCF
Gas remaining in reservoir = 80 − 22.8 = 57.2 MMMSCF
Oil remaining in reservoir $(N - N_p)$ = 100 − 20 = 80 MMSTB

Let us first consider repressuring of the reservoir to 4500 psia with a gas similar to the produced gas. Then at 4500 psia, the reservoir oil saturation would be

$$S_o = (N - N_p)\, B_o/NB_{oi} = \frac{80 \times 1.55}{100 \times 1.44} = 0.886$$

Also the total gas in the reservoir would then be

$$S_g\, NB_{oi}/B_g + (N - N_p)\, R_s = 0.\,114 \times 140/0.000551 + 80\,(1100)$$

$$= 117 \text{ MMMSCF}$$

Therefore, additional gas required is 117 − 57.2 = 59.8 MMMSCF to be injected in five years at an average rate of 33 MMSCF/D.

However, since gas injection will continue for an additional eight years, the total gas to be injected over the entire 13 years period is $(59.8)\,\dfrac{13}{5}= 155$

MMMSCF. This volume of gas has to be purchased and would cost $4.65 MM ($0.03/MSCF) but at least half of it would be later recovered, so the actual cost of the injected dry gas is $2.33 MM.

By the ideal isothermal formula, the horsepower required to compress available gas from 500 to 5000 psia, is

$$HP = \frac{33 \times 10^6 \times 14.7 \times 144 \times 2.3 \log \frac{5000}{300}}{1440 \times 33000} = 3360$$

and therefore the cost of the compressor plant is

3360 x \$300/HP = \$1.0 MM

From the ternary diagram it was established that for the given reservoir oil composition, miscibility is possible at the condition of 4500 psia and 250° F only by blending the dry gas with 0.2 barrel LPG per MSCF dry gas. It was further estimated that the "mixing zone" from "dry" gas to original oil composition should be 10 percent of the total reservoir hydrocarbon pore volume and that the blend in this mixing zone should contain one-half LPG volume. Therefore, 140 x 0.10 x 0.5 = 7 MMRB of LPG must be bought. If half of this is later recovered, the cost is \$10.5 MM (\$3.00/barrel).

On the basis of the permeability profile ($V = 0.50$), Figure 4.15 gives $\widetilde{V} = 3$. Also, the calculated effective mobility ratio is $E = 1.88$ (for $\mu_o/\mu_g = 10$, **Equation 6**, section 4.6, and $M = \widetilde{V}E = 5.64$.

From **Equation 13**, section 4.6, the percentage oil recovery is 59% after injecting 1 pore volume. Assuming an areal sweep efficiency of 80 percent due to pattern, dip, mobility ratio and pore volume of through-put, the total oil recovery by miscible drive is

$$S_o \text{ (pore volume)}/B_o \text{ (.59) (.80)} = \frac{.886 \, (144) \, (.59) \, (.80)}{1.5} = 40 \text{ MMSTB}$$

An additional recovery of 2 MMSTB can be expected during blowdown.

The net profit for the miscible venture is therefore

$$(40 + 2) \times \$1.00/\text{barrel} - (\underset{\substack{\text{com-}\\\text{pressor}}}{1.0} + \underset{\text{LPG}}{10.5} + \underset{\substack{\text{dry}\\\text{gas}}}{2.33}) = \$28 \text{ MM}$$

compared to \$12 MM by natural depletion.

Note that with gas injection we have assumed a net profit of \$1.00/barrel of oil produced. This reduction of \$0.20 per barrel compared to natural depletion is attributed to the cost of such items as additional wells, lines, maintenance of compressor plant, interest and salvage.

It should be noted that although miscible displacement efficiency is theoretically high, unfavorable viscosity ratios may adversely affect the overall process to the extent that recovery efficiency may be of the same order as a straight waterflood. For this reason, several variations of the basic miscible drive have been field tested, all with a view to counteract adverse mobility ratio effects.

Some of the combinations that have been suggested and/or tested are[4]: (1) simultaneous water and solvent injection, (2) alternate water and solvent injection; the theoretical effect being that the addition of water lowers the effective solvent mobility in the swept region causing the solvent to contact more oil sand. This is true if water and solvent flow as a uniform mixture; and (3) waterflood prior to miscible flooding; in this way the water will push the oil out of the tight areas into the large pores where the solvent is more effective.

In any event, the miscible process is particularly suited to reservoirs with some dip where gravitational effects may attenuate the adverse viscosity forces; to volatile oil reservoirs with a condensate gas cap already consisting of a rich equilibrium gas; and finally to those reservoirs susceptible to water injection or which have a low injectivity.

4.7 Performance Prediction of Partially Miscible Drives

There are a number of field projects of gas injection where complete miscibility is not practical but where near miscible conditions are feasible. Such may be the case for both condensing or vaporizing gas drives; in the former perhaps for a lack of a sufficiently rich gas and in the latter because of the too-high pressures required. Since in these processes the injected gas is usually different, composition-wise, from that liberated in the reservoir, the operation is also referred to as non-equilibrium gas displacement.

Predicting reservoir performance under this type of gas injection program where the fluids are only partially miscible, requires that we consider the same factors that influence immiscible displacement plus the effects of interphase mass transfer between the injected gas and the reservoir oil, which results in a redistribution of composition and liquid and gas saturations.

In effect, four factors must be taken into account in this type of model: (1) the vaporization or condensation between components in the gas or oil, with a resulting change in oil volume, (2) the variation of oil viscosity, (3) the increase in oil volatility at high pressures, and (4) the actual displacement of the oil by the gas.

Some three or four models of non-equilibrium gas displacement have been presented in the literature, each of these being applicable to either the vaporizing or condensing gas drive process. Two of these models, one due to Attra[1] and the other to Welge et al.,[37] are both based on the basic Buckley-Leverett theory, but making the necessary adjustments for phase effects. We will describe Attra's model in some detail, making specific reference to the vaporizing gas drive process in volatile oil reservoirs. The extension to condensing gas drives is straightforward.

Essentially, Attra's non-equilibrium model is a refinement of earlier compositional material balance methods which were not successful because they grossly exaggerate the effects of phase behavior by assuming the reservoir to be a well-stirred tank (unicell) with uniform composition and phase properties throughout. In order to eliminate this too simplifying assumption, Attra segmented the reservoir into various cells and adapted the Buckley—Leverett linear displacement model to account for phase changes at and behind the displacement front. The cell concept allows for a better approximation of the actual interacting phenomena between the injected dry gas and the reservoir oil.

The basic assumptions involved in the model are: (1) no capillary and gravitational effects, (2) phase equilibrium exists in each cell at the end of each time step, (3) the flow properties of each cell are based on a calculated average gas saturation, (4) the system is at nearly constant pressure; declining reservoir pressure may be approximated by a series of constant pressure drops, (5) constant injection velocity, (6) no significant mixing occurs along the direction of flow, and (7) continuous gas injection.

The basic data requirements for the non-equilibrium gas displacement computation are quite extensive, more so than for either condensate or completely miscible studies. These include:

Ternary phase diagram indicating the equilibrium information on the injection fluid-resident fluid compositional paths.

Densities and viscosities of the liquids and gases existing in equilibrium, whose compositions are shown on the ternary diagram and result from gas-oil interaction.

Equilibrium vaporization constants (K-values) which permit determination of the equilibrium phase compositions through repeated contacts of the displacing gas and reservoir fluid. These equilibrium constants constitute the most important data required since they must reflect compositional effects as pressure and temperature are usually constant.

Relative permeabilities to gas and to liquid as a function of saturation.

We will now outline the calculating procedure for the displacement process as the gas front moves into each cell (see Figure 4.17). The reservoir may be divided into as few as ten cells.

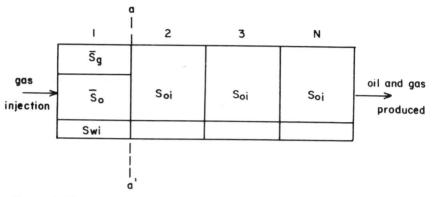

Figure 4.17. Displacement process as the gas front moves into each cell of the reservoir.

Step 1. Use the Welge tangent method (see section 3.1) to solve the fractional flow equation

$$f_g = \left\{ 1 + \frac{K_o}{K_g} \frac{\bar{\mu}_g}{\bar{\mu}_o} \right\}^{-1} \tag{1}$$

for an average value of gas saturation $\bar{S}_g$ behind the front when it reaches the position $a\text{-}a'$. This determines the volume of fluid displaced from cell 1 which is being invaded by the front.

Step 2. Based on this average saturation value, the residual oil and gas remaining in cell 1 are combined in proportion to their calculated volumes and are flashed at reservoir pressure and temperature to determine their equilibrium compositions.

The equilibrium flash condition is defined as

$$\Sigma_i \; \frac{z_i}{n_v + \left(\frac{1}{K_i} - 1 \right)} = 0 \tag{2}$$

and the molar composition of equilibrium oil and gas resulting from each flash is given by

$$\left. \begin{aligned} x_i &= \frac{z_i}{n_v (K_i - 1) + 1} \\[2mm] y_i &= K_i x_i \end{aligned} \right\} \tag{3}$$

n_v, the mols of gas per mol of fluid, is calculated from **Equation 2**; its value is then substituted in **Equation 3** to obtain the compositions (mol fraction) of oil x and of gas y. z is the composite fluid composition and K is the equilibrium constant.

Step 3. Knowing the compositions of the equilibrium gas and oil in cell 1, the viscosities of these phases are determined by the most convenient of the methods described by Katz *et al.*[17].

Step 4. These new viscosity values are used to recalculate the average gas saturation $\bar{S}_g$ (step 1). The test is made to compare this current value of the average displacing gas saturation with the original value from step 1. If agreement is not obtained, the new gas saturation and viscosities are used to adjust the residual volumes of fluid in cell 1 and steps 2 through 4 are repeated.

At the end of step 4, the gas front is at $a - a'$ (Figure 4.17) and the volume of fluid displaced from cell 1 has been calculated considering phase effects on the displacement mechanism, through the oil and gas viscosities. We must now concern ourselves with phase effects behind the front and how they affect the re-distribution of fluid compositions and saturations to satisfy phase equilibrium. We therefore

Step 5. Combine the residual oil and gas volumes behind the front and flash it at reservoir pressure and temperature. New fluid volumes are calculated at the new equilibrium compositions and compared to the previous cell volumes at the end of step 4.

Step 6. If agreement is not obtained between these values, the average gas saturation behind the front is adjusted to reflect the phase exchange and step 5 is repeated.

Step 7. When agreement is reached in step 6, the volume of fluids produced from the cell is flashed through surface separator conditions to determine gas-oil ratio, surface recovery and producing compositions.

Step 8. The entire procedure is repeated for succeeding time steps as the gas front moves into cells 2, 3, 4, ... N. Steps 1 and 4 are omitted for all cells behind the front since displacement behind the front is dependent only on two-phase flow and phase behavior.

In order to determine the relative volumes of oil and gas, R flowing behind and ahead of the front, the following equation is used

$$R = \frac{K_g}{K_o} \frac{\bar{\mu}_o}{\bar{\mu}_g} \tag{4}$$

The composition and volumes of fluid in each cell i at the beginning of one displacement step are related according to

$$z_i = \Sigma \frac{(1) + (2) + (3) + (4)}{\text{(mol of displaced liquid and gas +}} \tag{5}$$
$$\text{residual liquid and gas)}$$

where

(1) = (mol of displaced gas) $_{i-1}$ $(y)_{i-1}$
(2) = (mol of residual gas) $_i$ $(y)_i$
(3) = (mol displaced liquid) $_{i-1}$ $(x)_{i-1}$
(4) = (mol of residual liquid) $_i$ $(x)_i$

The procedure is now complete.

An idea of the importance of considering phase effects in gas injection operations in volatile crude oils, can be appreciated in the performance curves of Figure 4.18. These curves, computed by Attra, show cumulative stock tank oil recovery versus producing GOR for the actual field performance, using the

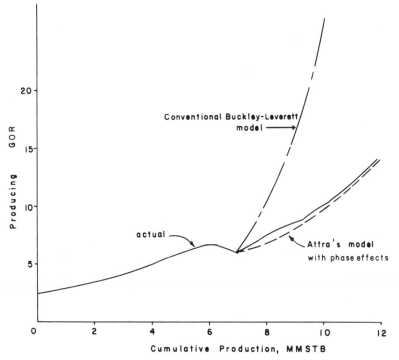

Figure 4.18. Field performance of dry gas injection program (after Attra[1]).

conventional Buckley-Leverett method, and considering phase effects. Table 8 illustrates the change of reservoir oil composition, resulting from the vaporizing effects of the injected gas. Note that the oil becomes stripped of the intermediates and the gravity of the produced crude changed from 50° to 37° API over a three year period. This also indicates that, because of phase behavior, field separator conditions must vary periodically so as to maximize stock tank recovery.

This stripping action of the gas, which operates progressively on the higher molecular weight components, evidently affects the composition of the C_7^+ fraction as gas cycling continues. For this reason particular care must be taken in establishing K-values for this fraction. Recently, Cook et al.[5] have presented correlations for estimating the K-values of the C_7^+ fraction together with a simple method for calculating which conditions of pressure, temperature, and type of oil are favorable for maximum vaporization.

Table 8. Effect of Gas Injection on Reservoir Oil Composition, Constant Pressure[1]

component	Mol Percent	
	original	after 3 years injection
C_1	52.0	44.0
C_2	6.3	4.9
C_3	5.3	1.5
C_4	6.5	0.2
C_5	3.6	0.1
C_{6+}	3.4	0.1
C_7^+	22.9	49.2
	100.0	100.0
API gravity	50	37

At this point, the reader may recall that in section 3.1 we presented modifications of the Buckley-Leverett displacement equations to allow for dilution of the reservoir oil caused by solution of gas and change of viscosity. Since these are the two principal mechanisms by which oil recovery is increased in the absence of complete miscibility, it is logical to enquire into the validity of that Buckley-Leverett simplified treatment.

In effect, the modification proposed in the previous chapter is based on the assumption that the reservoir oil becomes instantaneously saturated at the front by solution of the injected gas. It moreover assumes that no mass transfer occurs behind the front. These equilibrium assumptions are exact only if the saturated oil resulting from solution of gas has a composition which could exist in equilibrium with the gas; in other words, equilibrium gas injection. For this reason, it has been found[27] that the instantaneous approximation appreciably underestimates oil recovery from either vaporizing or condensing gas drive processes.

Finally, it should be pointed out that the cell principle suggested by Attra is not applicable to a slug displacement process because it introduces a smearing of the compositional front. In the case of slug injection projects (enriched gas or CO_2) under partial miscible conditions, a satisfactory procedure for calculating performance was proposed by de Nevers[6]. Again this method is based on the Buckley-Leverett model plus phase effects. Correlations for predicting solubility, swelling and viscosity behavior of CO_2-crude oil systems, as required by de Nevers' model, have been recently published by Simon and Graue[32].

4.N Closure

In this chapter we reviewed, condensed and interpreted a great deal of information from the literature concerning the miscible and partially miscible displacement of reservoir oil by gas injection. The object here was to provide the reader with an appraisal of the "state of the art" since the development of the knowledge of the behavior of the flow of miscible fluids in porous media is still in its early stage.

After discussing the solution theory of the better known vaporizing and condensing gas drive processes, the pertinent diffusion-convection models describing flow of homogeneous fluids in porous media are developed. Next, the LPG slug process is treated and various methods for determining the minimum slug size are examined critically. In particular, we review Perrine's method for slug size optimization under adverse viscosity and density differences of the system's fluids.

Finally, ample discussion is devoted to the more commonly used phenomenological models for predicting project performance under complete and partial miscible conditions. In this respect, the methods of Koval and Attra are favored mainly because they have been shown to be reliable, and also because they are built around more familiar models, such as those of Buckley-Leverett and Dietz, common to immiscible displacement theory. This provides more continuity in thought for the reader since the carry-over of ideas from one process to the other is greatly enhanced and simplified.

As in the previous chapters, several example problems are worked to facilitate an understanding of the theory presented.

Questions

1. a) Given the following data for the system H_2O – Alcohol – Benzene at 25°C. Construct the ternary phase diagram and draw in the tie lines.

Layer 1:
Wt. % C_6H_6:   1.3   9.2 20.0 30.0 40.0 60.0 80.0 95.0
Wt. % C_2H_5OH: 38.7 50.8 52.3 49.5 44.8 33.9 17.7   4.8

Layer 2:
Wt. % H_2O: —   —   3.2   5.0   6.5 13.5 34.0 65.5

b) Sketch a 3-D ternary diagram with pressure as the vertical parameter. Label the diagram completely.

2. Derive **Equation 1**, section 4.4, from the fundamental considerations of continuity and Fick's first diffusion law.

3. Obtain the solution of **Equation 5**, section 4.4, as given by **Equation 7**, section 4.4, by assuming that the coordinate system moves with velocity u_x. Note that under this assumption, real x is $u_x t$ less than the moving x' and therefore use the transformation $x' = x - u_x t$.

4. The description of miscible displacement requires a diffusion type equation for solvent continuity, **Equation 1**, section 4.4, an equation for the continuity of velocity $u_x + u_y + u_z = 0$ and Darcy's law for the equation of motion

$$\vec{u} = - \frac{K}{\mu(C)} \nabla (p + gp(C)z)$$

Obtain the following dimensionless groups

$$f \left\{ \frac{\mu_o}{\mu_s}, \frac{q\mu}{KL^2_g \Delta p_{o-s}}, \frac{q}{LD_e} \right\}$$

using the methods of inspectional analysis of section 3.11.

5. Derive **Equation 25**, section 4.4.

6. Show that the mixing zone length, for the case where the concentration is measured as a function of time at a fixed position in a linear system, is

$$\tilde{\Delta x} = (\sqrt{t_{90}} + \sqrt{t_{10}}) \sqrt{3.28 D_e}$$

where t_{90} and t_{10} are times when the concentration at the fixed position is 90 and 10 percent, respectively.

7. For very slow motion ($\bar{u} = 0$), **Equation 3**, section 4.5 abbreviates to

$$\tilde{C}(\tau) = \tfrac{1}{2}\left(erfc\ \frac{1}{\sqrt{\tau}} - erfc\ \frac{1}{\sqrt{\tau - \tau_1}} \right)$$

where

$$\tau = \frac{4D_e t}{x^2}, \qquad \tau_1 = \frac{4D_e t_1}{x^2}$$

a) Plot this equation as $\tilde{C}$ versus τ with $\tau_1 = \tfrac{1}{2}, 1, 3, 5$ as parameters and trace the locus $\tilde{C}^* = erfc\ 1/\sqrt{\tau}$. Discuss the propagation of the solvent slug as it moves through the system. Note that the convective term that was neglected does not distort the curves, since it only introduces a delay in the wave propagation.

b) Using the abbreviated form of **Equation 3**, section 4.5, given above, show that as τ_1 becomes small (x, large), the concentration $\tilde{C}(x, \tau)$ becomes proportional to τ_1. In other words

$$\lim_{\tau_1 \to 0} \left\{ erfc\ \frac{1}{\sqrt{\tau}} - erfc\ \frac{1}{\sqrt{\tau - \tau_1}} \right\} = \tau_1 \frac{d}{d\tau}\left(erfc\ \frac{1}{\sqrt{\tau}} \right)$$

8. Comment on the statement " . . . during stable miscible flow $k_o = k_s = k$ ". If the sand is water wet and the water saturation connate, then what value of the permeability is most appropriate?

Symbols

B_g, B_o	formation volume factors of gas and oil, vol/vol.
C	tracer or solvent concentration, mass of tracer/volume of solution.
$\tilde{C}$	relative solvent concentration, C/C_o, dimensionless
$\tilde{C}^*$	relative solvent concentration at minimum miscible conditions, dimensionless
C_A	concentration of solvent A in a binary mixture, mol fraction
C_o	solvent concentration at well bore, mass of solvent/volume of solution
$\bar{C}$	average relative solvent concentration, dimensionless

$\widetilde{C}_g$	relative phase volume concentration of gas, dimensionless
D'	intrinsic dispersivity coefficient, cm
D_e	effective dispersion coefficient, cm^2/sec.
D_o	molecular diffusion coefficient, cm^2/sec.
$D_o{}'$	apparent molecular diffusion coefficient, $= D_o/\tau$, cm^2/sec.
D_x	axial dispersion coefficient, cm^2/sec.
d_p	mean grain diameter, cm.
E	effective viscosity ratio defined by **Equation 6**, section 4.6
erf	error function, $= 1 - erfc$.
$erfc$	complementary error function defined by **Equation 4**, section 4.4
f_g, f_o, f_s	fractional flow of gas, oil and solvent, respectively, dimensionless
g	gravitational constant, atm-cm/gm.
g_1	function defined by **Equation 8**, section 4.5
h	formation thickness, cm.
i	index or initial condition
k	absolute permeability, darcy
k_g, k_o, k_s	effective permeability to gas, oil and solvent, respectively, darcy
K	phase equilibrium constant, dimensionless
L	length of porous medium, ft
M	$= \widetilde{V}E$, dimensionless
$\widetilde{n}$	Perrine's perturbation permeability variation property, dimensionless
n_A	mole fraction of component A in the binary mixture, dimensionless
n_v	mols of gas per mol of liquid, dimensionless
N	original oil in place, STB
N_p	cumulative oil produced, STB
$\widetilde{N}_p$	cumulative oil produced, pore volume
N_A	molar flux of component A, gm-moles/cm^2 - sec
p	pressure, atm
P_e	$= \bar{u}d_p\,\epsilon\,/D_o$, Peclet number, dimensionless
q^*	well injection rate per net sand thickness, **Equation 17**, section 4.4
q_w	injection rate at well, cc/sec

Q_i	cumulative volume of fluid injected, pore volumes
r	coordinate distance, cm
r_o	radius of capillary tube, cm
R_s	solution gas-oil ratio, SCF/STB
$\Delta \tilde{r}_i$	initial length of the undiluted solvent slug, cm
S, S_o, S_g	saturation of solvent, oil and gas, respectively, dimensionless
S_{oi}, S_{wi}	initial oil and water saturations, respectively, dimensionless
$\tilde{S}_o, \tilde{S}_g$	average oil and gas saturations behind the injection front, dimensionless
t	time, sec
t^*	time to inject a graded solvent, sec
t_1, t'	time of injection of pure solvent, sec
T	temperature, °R
u	Darcy velocity, cm/sec
$\bar{u}$	average pore velocity, cm/sec
u_c	critical Darcy velocity of instability, cm/sec
u_r, u_x	velocity components in r and x-directions, cm/sec
V	Dykstra-Parsons permeability variation factor, dimensionless
V	Koval's heterogeneity factor, dimensionless
V_A	volume of component A evaporated, cc
w	integration variable
x	coordinate distance, cm
$x_{.9}, x_{.1}$	position of the 90 and 10-percent solvent concentrations, cm
$\tilde{\Delta x}$	length of the 90-10 percent mixing zone, cm
x_i	mol fraction of oil component i
y_i	mol fraction of gas component i
z_i	composite fluid composition of component i, mol fraction
α	dip angle of formation, degrees
β	tilt angle of solvent-oil "interface", degrees
ϵ	inhomogeneity or packing factor, dimensionless
μ_g, μ_o, μ_s	pure gas, oil and solvent viscosities, respectively, cps
$\bar{\mu}_o, \bar{\mu}_g$	oil and gas viscosities adjusted for phase effects, cps
$\tilde{\mu}$	viscosity of fluid mixture, cps
$\Delta \mu_{o\text{-}s} = \mu_o - \mu_s$	
ρ_o, ρ_s	pure oil and pure solvent densities, respectively, gm/cc
$\tilde{\rho}$	density of fluid mixture, gm/cc
τ	effective rock tortuosity, dimensionless

References

[1] Attra, H.D. "Nonequilibrium Gas Displacement Calculations." *Trans. AIME* (1961) 222, II-130.

[2] Bear, J., Zaslavsky, D. and Irmay, S. "Physical Principles of Water Percolation and Seepage." *UNESCO Arid Zone Research* – XXIX (1968).

[3] Brinkley, T. W. and Dorsey, J. B. "Performance Review of Shoats Creek Unit Vaporizing Gas-Drive Project." *Trans. AIME* (1968) 243, I-416.

[4] Cone, C. "Case History of the University Block 9 (Wolcamp) Field-A Gas-Water Injection Secondary Recovery Project." *Jour. Petroleum Tech.*, December 1970, Pag. 1485.

[5] Cook, A. B., Walker, C. J. and Spencer, G. "Realistic K-Valves of C_7^+ Hydrocarbons for Calculating Oil Vaporization During Gas Cycling at High Pressures." *Trans. AIME* (1969) 246, I-901.

[6] de Nevers, N. H. "A Calculation Method for Carbonated Water Flooding." *Trans. AIME* (1964) 231, II-9.

[7] Dodson, C. R., Goodwill, D. and Mayer, E. H. "Application of Laboratory PVT Data to Reservoir Engineering Problems." *Trans. AIME* (1953) 198, 287.

[8] Dougherty, E. L. "Mathematical Model of an Unstable Miscible Displacement." *Trans. AIME* (1963) 228, II-155.

[9] Dykstra, H. and Mueller, T. D. "Calculation of Phase Compositions and Properties for Lean- or Enriched-Gas Drive." *Trans. AIME* (1965) 234, II-239.

[10] Elliot, J. K. and Johnson, H. C. "Control of Injection Gas Composition in Enriched Gas-Drive Projects." *Trans. AIME* (1959) 216, 398.

[11] Fitch, R. A. and Griffith, J. D. "Experimental and Calculated Performance of Miscible Floods in Stratified Reservoirs." *Trans. AIME* (1964) 231, I-1289.

[12] Grives, R. B. and Thodos, G. "The Cricondentherm and Cricondenbar Pressures of Multicomponent Hydrocarbon Mixtures." *Trans. AIME* (1964) 231, II-240.

[13] Hall, H. N. and Geffen, T. M. "A Laboratory Study of Solvent Flooding." *Trans. AIME* (1957) 210, 48.

[14] Hoopes, J. A. and Harleman, D. R. "Dispersion in Radial Flow from a Recharge Well." *Jour. of Geophysical Research,* Vol. 72, No. 14, July 15, 1967, p. 3595.

[15] Jacoby, R. H. and Berry, V. J. "A Method for Predicting Depletion Performance of a Reservoir Producing Volatile Crude Oil." *Trans. AIME* (1957) 210, 27.

[16] Katz, D. L. "Possibility of Cycling Deep Depleted Oil Reservoir after Compression to a Single Phase." *Trans. AIME* (1952) 195, 172.

[17] Katz, D., Cornell, D., Kobayashi, R., Poettmann, F., Vary, J., Elenbaas, J. and Weinaug, C. *Handbook of Natural Gas Engineering."* New York: McGraw-Hill Book Co., 1959.

[18] Koonce, K. T. and Blackwell, R. J. "Idealized Behavior of Solvent Banks in Stratified Reservoirs." *Trans. AIME* (1965) 234, II-318.

[19] Koval, E. J. "A Method for Predicting the Performance of Unstable Miscible Displacement of Heterogeneous Media." *Trans. AIME* (1963) 228, II-145.

[20] Kyle, C. and Perrine, R. L. "Experimental Studies of Miscible Displacement Instability." *Trans. AIME* (1965) 234, II-189.

[21] Legatski, M. W. and Katz, D. L. "Dispersion Coefficients for Gases Flowing in Consolidated Porous Media." *Trans AIME* (1967) 240, II-43.

[22] Lohrenz, J., Clark, G. and Francis. R. "A Compositional Material Balance for Combination Drive Reservoirs with Gas and Water Injection." *Trans. AIME* (1963) 228, I-1233.

[23] Menzie, D. and Neilsen, R. F. "A Study of the Vaporization of Crude Oil by Carbon Dioxide Repressuring." *Trans. AIME* (1963) 228, I-1247.

[24] Ogata, A. and Banks, R. "A Solution of the Differential Equation of Longitudinal Dispersion in Porous Media." U.S. Geological Survey Professional Paper 411-A Washington (1961).

25 Perkins, T. K. and Johnston, D. C. "A Review of Diffusion and Dispersion in Porous Media." *Trans. AIME* (1963) 228, II-63.

26 Perrine, R. L. "Stability Theory and Use to Optimize Solvent Recovery of Oil." *Trans. AIME* (1961) 222, II-9.

27 Price, H. S. and Donahue, D.A.T. "Isothermal Displacement Processes with Interphase Mass Transfer." *Trans. AIME* (1967) 240, II-205.

28 Raimondi, P., Gardner, G. and Petrick, C. "Effect of Pore Structure and Molecular Diffusion on the Mixing of Miscible Liquids Flowing in Porous Media." Preprint No. 43 presented at the A.I. Ch. E.-SPE Joint Symposium in San Francisco (Dec. 6-9, 1959).

29 Reid, R. C. and Sherwood, T.K. *The Properties of Gases and Liquids,* New York: McGraw-Hill Inc., 1966.

30 Rowe, A. and Silberberg, I.H. "Prediction of the Phase Behavior Generated by the Enriched-Gas Drive Process." *Trans. AIME* (1965) 234, II-160.

31 Rutherford, W. M. Miscibility Relationships in the Displacement of Oil by Light Hydrocarbons, *Trans. AIME* (1962) 225, II-340.

32 Simon, R. and Graue, D. J. "Generalized Correlations for Predicting Solublity, Swelling and Viscosity Behavior of CO_2-Crude Oil System." *Trans. AIME* (1965) 234, I-102.

33 Slobod, R.L. and Koch, H.A. High Pressure Gas Injection-Mechanism of Recovery Increase, *API Drill, and Prod. Prac.* (1953), 82.

34 Stone, H. and Crump, J.S. "The Effect of Gas Composition Upon Oil Recovery by Gas Drive." *Trans. AIME* (1965) 207, 105.

35 Taylor, G. I. "Dispersion of Soluble Matter in Solvent Flowing Slowly through a Tube." *Proc. Roy. Soc.,* London A (1953) 219, 186.

36 Thompson, J. L. "A Laboratory Study of an Improved Water-Driven LPG Slug Process." *Trans. AIME* (1967) 240, I-319.

37 Welge, H. J., Johnson, E.F., Ewing, S.P. and Brinkman, F.H. "The Linear Displacement of Oil from Porous Media by Enriched Gas." *Trans. AIME* (1961) 222, I-787.

Appendix A

Special Problem Sets*

Problem I: Material Balance Methods applied to Jones Pool.

1. Using the data on the Jones Sand Pool, Chapter 1, plot the proper quantities from the general material balance equation which would give a linear plot· if there were no water drive and no initial gas cap. Estimate the initial stock tank oil N in place by extrapolation to zero production. How does the plot indicate a water drive?

2. Take the value of m (initial ratio of gas cap volume to oil zone volume) from the geologic data given, assume the simple Schilthuis invasion integral and, by several trials for the invasion constant, linearize the material balance plot, determining the initial oil in place and the constant B_1 of the Schilthuis invasion formula.

3. Using the material balance in the above linear form, take the group of equations (omitting data prior to 1940) and obtain "best" values for N and B_1 by the method of averages or the method of least squares. (Note that, in the method of averages, only the earlier data must be in one group and only the later data in the other.)

4. From the initial oil in place, the sand volumes, and other necessary data, calculate the average initial water saturation assuming that water is also present in the gas cap, at the same saturation as in the oil zone.

*Based on case histories of Chapter 1.

5. Assume primary depletion, without gas injection, over a ten year period, and assume that the pressure decreases linearly with time from 3520 to 320 psia, evenly distributed. Using the water invasion constant and the value of m from above, and assuming that 75% of all the original gas (free and dissolved) is produced and no water produced, calculate the cumulative production and per cent recovery after the ten years.

6. Also calculate the residual saturation in the non-invaded zone, assuming $\overline{S}_o = 0.30$, $\overline{S}_g = 0.05$, in the water invaded portion and the rest of the oil evenly distributed over the non-invaded oil zone. The formation volume factors are assumed to correspond to 320 psia in both invaded and non-invaded zones, as required by the material balance.

7. Assume a production-time schedule, such that the pressure declines 150 psi per year, and half the invaded water up to any time is produced. The Schilthuis invasion law is obeyed, with the constant as determined above. Gas injection is such that the cumulative gas/oil ratio is 1000 SCF/STB at any time. Calculate the production after four, eight, 12, 16 and 20 years, that is, at pressures 2920, 2320, 1720, 1120, and 520 psia.

8. From a plot of the previous cumulative productions versus time, estimate the production rate at the different times. Also calculate the oil saturation in the uninvaded part at 520 psia, using the assumptions of Problem I.6 except that the formation volume factors are taken at 520 psia. (Note that the net water invasion is half of that given by the invasion integral).

9. Take the relative permeability ratio as $k_g/k_o = 3 (S_g/S_o)^3$ and calculate the producing gas-oil ratio at 520 psia and the gas production rate, SCF/day. If the net gas-oil ratio (after injection) is 1000 SCF/STB, what compressor horse power is required to inject the rest of the production at 2000 psia, assuming 200 psia compressor intake? Take the viscosity ratio as $\mu_o/\mu_g = 60$.

10. Complete the W_e column in the production table using the steady state invasion formula and the constant calculated previously, taking an average pressure decline (from the original) over each two year period. Also calculate the final W_e by material balance and compare. Using the latter value, calculate the final oil and gas saturation, averaged over the original oil zone. (If gas saturation is calculated to be negative, call it zero).

11. If the operators had decided to operate this field as a natural water drive after July, 1941, when the pressure had dropped to 1542 psia, how long would it take to deplete the field, assuming that gas above solution GOR

was injected, and the production rate (reservoir oil) such as to balance the water invasion rate? Water production is assumed negligible, that is, wells are shut in as the water reaches them. Residual saturations are as given in Problem I.6, with formation volume factors taken at 1542 psia.

12.   From Problem I.11, if it is apparent that the life of the field would have been too long, a faster rate would involve a choice between depletion (calculated in Problem I.5) or maintaining pressure by water and/or gas injection. Using the information on displacement efficiency given in Chapter 3, discuss semiquantitatively the reservoir mechanics and economic factors to be considered in the various choices, and the probable reasons for the operators' choice of both gas and water injection. Note answers to previous parts of this problem and the data of Table 1. Why was gas injection abandoned toward the end?

Answers

1.   ca. 100 MM STB
2.   and p. ca. 100 MM STB; ca. 960 bbl yr^{-1} psi^{-1}
4.   0.38
5.   35 MM STB or 35% (15.4 MM bbl water invaded)
6.   0.335
7.   4.5, 12.0, 23.3, 38.6, 59.0 STB
8.   ca. 4, 7, 9, 11, 19 M STB/D; 0.20
9.   32 MSCF/STB; 600 MM SCF/D; 60,000 HP
       (Problems 7, 8 and 9 are hypothetical)
10.   23.4, 27.6, 31.7, 35.7, 39.2, 42.4 MM B; 45 MM B; 0.20 and 0.10
11.   25 years

Problem II: Complete Dispersion, Partial Reinjection and Concept of "Conformance," Frio Sand

Assumptions

Injection into the cap, and also cap expansion assumed completely dispersed in the initial oil leg. Production uniformly dispersed over original oil leg. Also use

$$k_g/k_o = 2.36 \left\{ (0.76 - S_o)/S_o \right\}^3 \qquad \text{A-1}$$

1.   From the Muskat equation, calculate the pressure versus production and pressure versus gas/oil ratio curves. Prepare the necessary values and

expressions for the quantities in the equation and in auxiliary equations such as the saturation equation, for no gas return and for 70 per cent gas return, assuming that a machine program is available. If the program is not available, the instructor will provide alternatives.

2. Using an integrated form of the material balance, calculate the production at 1200, 800, and 300 psia: a) assuming injection of all produced gas, and b) assuming injection so that the average net gas/oil ratio over the time period is 500 from 1590 to 1200 psia, 800 from 1200 to 800 psia, and 1000 from 800 to 300 psia. Net gas/oil ratio = gas not reinjected/oil produced. (In the first case it never drops to 300 psia, hence try 1000 psia).

3. Take one of the cases in which part of the gas is reinjected and set up a schedule of production rate versus time for 20-year depletion, with abandonment at 300 psia, if production rate at any time is proportional to the reservoir pressure. (Higher relative permeability in the early stages compensates for the fewer number of wells). Calculate the necessary horse power of the compressor station for the time when the most power is needed, assuming compressor intake 200 psi less, and compressor output 200 pounds more, than the average field pressure at that time. Use the idealized isothermal formula at $60°F$, standard for measuring gas-oil ratio. Taking $k = 194$ md, 10 foot well penetration, $r_w = 0.3$ ft, $r_e = 10,000$ feet, $p_e - p_w = 200$, is one injection well enough?

4. With costs and other economic data supplied by the instructor, determine whether a reinjection program is worth while. Pumping is required in all cases.

5. The concept of "conformance" assumes that in a fraction e (the "conformance factor") of the reservoir there is complete "conformance", that is, all the injected gas is completely dispersed in this zone, while the non-conformable fraction $(1 - e)$ behaves as if it were under solution gas drive only. By some strange mechanism, the whole reservoir has the same pressure at the same time, but the value of e is constant. Since some of the reinjected gas comes from the nonconformable zone but enters only the conformable zone, the history of the reservoir cannot be calculated by simply combining the two parts of Problem II.1. However, the material balance assumes only uniform pressure and is not dependent on saturation distributions. Hence, if reinjection is expressed as a fraction of total produced gas, it is a matter of finding an expression for the net k_g/k_o ratio which governs the producing gas-oil ratio, as in cases where frontal drive or segregation mechanisms are combined with pressure depletion (see Problem III). If the oil saturation for the non-conforming zone S_{on} at any

pressure is known from a depletion drive calculation and S_o is the average for the reservoir by material balance, then S_{oc}, the saturation in the conforming zone, is $S_{oc} = S_o - (1-e) S_{on}/e$, and hence k_g/k_o can be found for each zone. However, the effective k_g/k_o for the reservoir depends on the number of wells and their relative productivities in each zone. For the sake of calculation, assume the net k_g/k_o to be $(1-e)$ $(k_g/k_o)_n + e (k_g/k_o)_c$ at any time and $e = 0.6$, and calculate the production versus pressure and GOR versus pressure curves, by modifying the program used in Problem II.1.

Answers

2.   (a) $N_p/N = 0.36$ at 1200 psia. $N_p/N = 0.75$ at 1000 psia, this being ultimate after infinite time and recycling if $k_{or} = 0$ at $N_p/N = 0.75$.
     (b) $N_p/N = 0.14$ at 1200 psia, 0.28 at 800, 0.46 at 300.

3.   Using the values from Problem II.2, take N_p/Nas linear in p. If $N_p^* = N_p/N$, simple calculus gives $0.082 (0.57 - N_p^*)$ for dN_p^*/dt, with t in years and $t(\text{yr}) = (20/1.64) \ln 0.57/(.57-N_p^*)$. Daily production $= N (dN_p^*/dt)/365$.

p	N_p^*	t (yr)	dN_p^* / dt	STB/D
1590	0	0	0.0466	2800
1310	.1	2.3	.0385	2300
1030	.2	5.3	.0303	1800
750	.3	9.1	.0221	1300
485	.4	14.7	.0139	830
300	.46	20	.0090	540
(0	.57	∞	0	0)

Because of the rapid change of relative permeability ratio with saturation, it appears that the most gas is produced near abandonment. From Problem II.2, at $N_p/N = 0.46$, $S_o = 0.38$, $S_g = 0.38$, $k_g/k_o = 2.36$, $\mu_o/\mu_g = 64$ at 300 psia, $B_o/B_g = 0.01$, whence 15000-1000 SCF/B or 540·14000 SCF/D are reinjected, or 7.5 MMSCF/D. At a compression ratio of 500/100, this requires 550 HP. At $p_e = 300$, $p_w = 500$, $k_{rg} = 1$, an injection well can take 2.7 MMSCF/D, so three are required unless injection pressure is higher, requiring higher HP.

Problem III: Complete Pressure Maintenance, Production Strictly by Down Dip Displacement, Injection Into Cap, Based on Frio Sand

Assumptions

In addition to the assumptions in the problem title, assume up dip wells shut in as soon as the gas front arrives. The total field production rate is constant and such that all the oil recoverable by such operation is recovered in 10 years. Assume homogeneity and constant sand thickness. Pressure maintained above 1590 psia, the original reservoir pressure.

1.   Using the field of Problem II, construct the Buckley-Leverett-Welge diagram neglecting gravity (see table for calculations). What is the average residual oil saturation over the gas-invaded part of the oil leg, if wells are shut in at gas break-through? What is the theoretical per cent recovery of oil from the invaded part? Compare with the recovery by depletion methods, Problem II. What is the total production rate (STB/D), making all the previous assumptions?

Additional Data

S_g	S_o	$2.36\left(\dfrac{S_g}{S_o}\right)^3$	$\dfrac{k_g\,\mu_o}{k_o\,\mu_g}$		$\dfrac{k_g\,\mu_o\,B_o}{k_o\,\mu_g\,B_g}$		R		f_g	
			1200	1590	1200	1590	1200	1590	1200	1590
0.1	.66	.0082	.377	.312	265	252	550	570	.274	.238
.15	.61	.0349	1.605	1.325	1125	1068	1410	1387	.615	.570
.20	.56	.0843	3.87	3.20	2710	2580	2995	2900	.795	.762
.25	.51	.278	12.8	10.6	8975	8550	9260	8870	.928	.915
.30	.46	.654	30.1	24.8	21100	2000	21400	20300	.967	.962
.35	.41	1.46	67.4	55.5	47200	4475	47500	45100	.985	.982

2.   To get some idea of the contribution of the gravity term in the fractional flow formula, assume a circular inverted cone-shaped structure, 10 foot thickness, and the production rate just calculated. Take the cross-sectional area so that half of the total reservoir volume is on each side. From one or two points on the Welge diagram, is the gravity term significant?

3.   Neglecting produced gas, which is all solution gas under the present assumptions, how many SCF of gas (same properties as the reservoir gas) must be bought over the life of the project? If bought at 500 psia and compressed to 2,500 psia, estimate roughly the horse power requirements.

4.   If the field had been allowed to drop to 1200 psia, with no gas return (or gas return to give an average GOR of 500 SCF/STB, see Problem II) and then shut in while repressuring to 1800 psia, how many SCF of gas is required for the repressuring? As an approximation, assume the oil remaining in the reservoir to become saturated at 1590 psia as a result of the repressuring. From Problem III.3, how long will this take with the available compressor capacity?

5. Assume that essentially the whole reservoir can be swept with gas-oil ratios remaining at the solution gas value (at 1590) and that, after complete sweeping, the reservoir will be blown down to a low pressure with injection stopped. Estimate the additional oil recovery, STB, and the total gas, SCF, also the instantaneous gas-oil ratio at abandonment, SCF/STB. What is the total production, STB, just before blow-down, assuming the residual oil saturated at 1800 psia? (Hints: the total produced gas corrected to reservoir pressure at any time is, in pore volumes,

$$S_g \int_{p_1}^{p_2} dp/p = S_g \ \ln \ (p_1/p_2) \ \text{if} \ S_g \ \text{is a constant} \qquad \text{A-2}$$

and no gas comes out of solution.) This is not true, but, as an approximation, take $S_g = 0.25$ from Problem III.1 and $\ln (p_1/p_2) = \ln (1800/200)$, giving $0.25 \ \ln 9 = 0.55$. Considering the previous injection of 0.25 pore volume (at 1800 psia) and gas from solution, assume one pore volume total throughput for displacement calculations. The reciprocal of this (also unity) is the slope of the tangent to be drawn on the Welge diagram to get the average residual saturation. Note: Strictly speaking, the tangent for production before blow-down should be drawn to allow for saturating the original oil from 1590 to 1800 as in Problem IV.

6. What is the theoretical producing gas-oil ratio, SCF/STB, just after breakthrough of the gas-invaded region? Assuming the structure of Problem III.2, draw contour circles showing the original gas-oil interface (looking down on the field from high in the air), the "front" after six years' production, and the loci where gas-oil ratios would be 50,000 and 20,000 SCF/STB, if wells at those levels were momentarily opened after six years' production.

7. Assuming $r_e = 10{,}000$ feet, "effective" $r_w = 0.333$ feet, $k_{rg} = 1, p_e = 1600$ psia, $p_w = 24000$ psia, how many injection wells are needed (see Problem III.3)

8. If, after blowing down to 200 psia average pressure (Problem III.5), it is decided to continue production from the bottom wells and maintain that average pressure by resuming injection, how much more gas (SCF) would have to be injected before the gas-oil ratio reaches 50,000 SCF/STB? And how much more oil would be produced? With the available horse power and a compression ratio of 2 (200 psi intake and 400 psi discharge) how long would this take? With $p_e = 200$ and $p_w = 400$ and the other data same as Problem III.7, would some up dip wells have to be converted to

injection wells, and how many, if any, if the total compressor capacity is to be used?

Answers

1.   0.50; 34%; ca. 2000.

2.   No.

3.   Ca. 7 MMMSCF; ca. 1200 H.P.

4.   4.1 MMMSCF; six years.

5.   0.6 MMSTB; 16 MMMSCF; 2000 SCF/STB; 7.3 MMSTB.

7.   One.

Problem IV: Partial Reinjection Into Cap, With Allowance for Segregation Due to Displacement (Neglecting Gravity), Frio Sand

Assumptions

Assume injection into the cap of 70% of the produced gas, **after** the average field pressure has declined to 1200 psia. See Problem 11.2 (b) for production to this point with the GOR assumed constant at 500 SCF/STB. The gas expands down dip, displacing oil in accordance with displacement theory (neglecting gravity). Wells, if originally drilled high on structure, are shut in when the expanded gas cap reaches them, so that production is always from the shrinking oil zone. Gas does not penetrate into the shrinking oil zone as long as displacement theory allows for the formation of a "front". After that, V_{pinv} (inv = invaded) in the equations remains constant at its value just before that. Also

$$k_g/k_o = 2.36 \left\{ (0.76 - S_{oo})/S_{oo} \right\}^3 \qquad \text{A-3}$$

where S_{oo} is the average oil saturation in the producing (shrinking) oil zone. This is related to the average taken over the original oil zone, the S_o calculated by the Muskat material balance, by the relation

$$S_{oo} = (S_o - \bar{S}_{or} V_{pinv})/(1 - V_{pinv}) \qquad \text{A-4}$$

where $\bar{S}_{or}$ is the average residual oil saturation by displacement theory in the region invaded by the expanding cap and V_{pinv} is the pore volume invaded, as fraction of the original oil zone pore volume.

$$V_{pinv} = V_{tp}/\overline{S}_{gr} \qquad\qquad \text{A-5}$$

where

$$V_{tp} = Q_{inj} B_g + m(B_g/B_{gi} - 1) \qquad\qquad \text{A-6}$$

V_{tp} = volume gas throughput at reservoir conditions; $Q_{inj} = \Sigma \Delta Q_{inj}$, and ΔQ_{inj} = $rR \, \Delta N_p = rR \, \Delta S_o/B_o$, r = fraction injected.

If gas dissolves or comes out of solution in the invaded region, the following equation must be satisfied by graphical trial and error on the Welge diagram

$$\overline{S}_g - S_{gi} + (1 - S_w - \overline{S}_g)(R_s - R_{si}) \, B_g/B_o = 1/f_g' \qquad\qquad \text{A-7}$$

where f_g' is the slope of the tangent from $\overline{S}_g$ on the $f_g = 1$ line and R_s and R_{si} are the ambient and initial solubilities.

Note that the above equations would serve no purpose if definite net gas/oil ratios were assumed, Problem II.2, material balance being independent of saturation distributions. Recall, however, that all material balance methods assume uniform pressure and phase equilibrium at any stage.

1. Assuming that a program is available, calculate the pressure versus production and pressure versus gas-oil ratio curves for tthe above assumptions. Compare with calculations of Problems II and III.

2. From the Welge diagram, what is the maximum gas saturation in the "shrunken" oil zone that would allow "front" formation?

Problem V: Displacement Theory, including Coverage Factor, Applied to Brookhaven Field

1. Assuming that this field remained undersaturated during the initial pressure depletion to 2000 psi, calculate the effective oil compressibility, psi^{-1}, which includes the effect of water and the formation.

2. Check the figure given (180 MM) for the initial stock tank oil in place by volumetric methods.

3. Using the viscosities of the oil and gas saturated at the maintenance pressure 2850 psia, and the relative permeability ratios given, construct the Welge diagram neglecting gravity and the dissolving of injected gas into the undersaturated oil. From the intercept of the tangent, $\overline{S}_g$, what is the average oil saturation in the gas invaded region before "breakthrough" and the per cent recovery at "gas breakthrough"? (Estimated recovery by depletion was 20%).

4.   To allow for gas dissolving in the residual oil, up to the saturation point, use **Equation A-7** of Problem IV. Take, R_s = 433 and B_o = 1.30. For the saturated oil, values are given in Chapter 1, B_o in the equation referring to the saturated oil. From the Welge diagram so constructed, what is the average oil saturation in the region behind the front before gas breakthrough, and the per cent recovery at breakthrough?

5.   Derive the expression for f_g in terms of R (producing gas-oil ratio), B_o, and B_g, neglecting production of solution gas.

6.   Neglecting solution gas produced, from the Welge diagram what is the residual oil saturation and the percent recovery if production is continued, after gas breakthrough, to a gas-oil ratio of 50,000 SCF/STB?

7.   Show the calculation of the "coverage factor" of 0.55 from the production and other data given. Assume B_o = 1.3 in the uninvaded and water invaded regions and B_o = 1.4 in the gas invaded region. The "coverage factor" is the ratio of the oil produced (after subtracting that due to water drive) to the expected production, according to calculated displacement efficiency (take S_g = 0.22) and the sand volume invaded by the gas drive.

Answers

1.   2.8 x 10^{-5} psi^{-1}

3.   0.40; 20%

4.   0.38; 36%

5. $(1 + \dfrac{B_o}{RB_g})^{-1}$

6.   0.32

Problem VI: Gravity Segregation as a Major Factor, Mile Six Pool

In Chapter 1, Mile Six Pool was cited as a case in which good advantage could be taken of gravity segregation. For a problem based on this field, some additional information, not given in Chapter 1, is needed. This includes: absolute permeability 0.3 darcy, hydrocarbon porosity 0.1625, and connate water saturation 0.35 of total porosity. Relative permeabilities and permeability ratios as functions of "effective gas saturation" S_{ge} (fraction of hydrocarbon porosity) are as follows:

S_{ge}:	.05	.1	.15	.2	.3	.4	.5	.6
k_{ro}:	.77	.59	.44	.34	.19	.10	.04	0
k_o/k_g:	∞	38.	8.8	3.1	.72	.21	.072	0

1. Construct the Welge diagram for down dip reservoir fluid flux of 20, 11.5, and 6 MRB/D, assuming a cross section perpendicular to flow of 1.24 million square feet, pressure being maintained by gas injection into the cap. (See Craft and Hawkins).

2. What is the average oil saturation and fraction of original oil in the invaded zone in each case?

3. What would be, approximately, the life of the field if, in each case, the reservoir oil production remained constant and corresponded to the above fluid flows?

4. Assuming pressure in each case, what economic factors and operational procedures would be involved in each case if the same ultimate production was to be obtained? (Note that the second case was the one corresponding to the actual history).

5. In the Welge diagram for the lowest rate, negative values of fractional gas flow are obtained at some of the lower gas saturations, implying counter-flow. Does this actually occur? Explain on the basis of the diagram.

6. Noting the presence of a water drive, give possible reasons why depletion or partial depletion combined with the water drive would have been less economical than the operation adopted.

7. What is the maximum possible oil recovery on the basis of the relative permeability relations? How would this be obtained, practically, considering the lower "breakthrough recoveries" in Problem VI.2?

Answers

2. 0.87, 0.65. 0.48, saturations based on hydrocarbon porosity.

3. one, five and 15 years.

7. 60 percent or 30 MMSTB.

Problem VII: Solvent Slug in a Pattern, Parks Field

1. Assuming no water invasion, calculate the initial oil in place by material balance. Note that the volumetric estimate was 23 MMSTB. From a correlation chart for reservoir oils, the gas solubility is estimated at 1900 SCF/STB at the initial bubble point, 3500 psia, and 1000 at 1900 psia. Take Z for the gas as 0.9. Neglect expansion of water and rock.

2. Assuming the 23 injection wells, what was the hydrocarbon pore volume hopefully to be displaced from each? What is the sand volume, area, and effective radius controlled by each? Assuming complete homogeneity and no gravity overriding or viscous fingering, what should be the theoretical minimum slug size for each well by the formula of Fitch and Griffith? Take the oil viscosity as 0.4, the LPG viscosity as 0.1, and the gas viscosity as 0.02, and assume the linear ratio $\Delta x/L$ of F. and G. to apply to the volumetric ratio in pattern flow, L being the effective radius in feet controlled by an injection well.

3. The vertical permeability profile was not given but, assuming three non-communicating layers of thicknesses two, four, and 14 feet and permeabilities, respectively, of 20, 5 and 1 md, what would be the vertical sweep efficiency at breakthrough by Stiles' method, taking the mobility ratio as unity? Give reasons why the actual vertical sweep efficiency was much higher, assuming the permeability profile to be correct.

4. Why is water injection with or following the gas expected to increase horizontal sweep efficiency?

Answers

1. 24 MMSTB

2. 1.5 MMbbl; 33 MMbbl or 185 MM   cu ft; 9.3 MM sq ft; 1800 ft; about 0.006 of hydrocarbon pore volume.

3. 18 per cent.

Problem VIII: Propane Slug Oriented by Gravity, Baskinton Field

1. Sketch the field from the description given, showing shape, dip, thicknesses, wells (including injection well) and lengths of semi-major and semi-minor axes of the elliptical area.

2.   Give a possible explanation as to why the cumulative gas/oil ratio for the primary period was less than the original gas solubility.

3.   Taking a gas-oil interface 3000 feet long parallel to the major axis where the formation thickness is 20 feet, what is the linear rate of advance (assumed in the bedding plane and in the direction of the minor axis), if the rate of volume increase behind the interface due to this advance is given by an oil recovery rate of 300 RB/D? (Note porosity and water saturation given).

If the advance of an interface due to gas injection continued down dip for a very long time without encountering wells, the interface would not only become horizontal but the tilt would eventually go beyond this and be down dip. While this is not likely here, what would be the tilt according to the Dietz formula for small dip Chapter 3, **Equation 12**, section 3.6? Take k in perms ($6.3 \cdot k$ in darcys), densities in lb/cu ft, u ($= v \phi S_o$) in cu ft/day, $g = 1/144$, relative permeabilities unity, viscosities 0.8 cp and 0.02 cp for oil and gas, and densities 0.7 and 0.1 compared to water.

What is the velocity component downward (vertically)? Calculate the "critical" (maximum) downward velocity to avoid viscous gravity fingering of the propane layer into the oil, assuming a sharp, horizontal boundary. Density of liquid propane 0.6 compared to water, viscosity 0.1 cp. (Note that this is dependent on a knowledge of viscosities and densities for field conditions, also on a sharp interface, which is soon dissipated by dispersion and mixing). Use Gardner's formula

$$u_c = \frac{k\,(\rho_1 - \rho_2)\,g}{\phi\,(\mu_1 - \mu_2)} \qquad\qquad \text{A-8}$$

with the units previously given.

Answers

1.   Semi-axes 2700 and 1350 feet.

3.   0.14 ft/day; $\tan^{-1}$ 0.34, which is more than the formation dip, hence "tonguing" of the gas phase would eventually take place if times and distances were great enough. 0.005 ft/da; 0.54 ft/da.

Problem IX: High Pressure Injection, Block 31 Field

1. Point A in the ternary diagram of Figure A.1, shows the composition of the crude in the reservoir after the depletion by liquid expansion of the undersaturated oil. From the diagram, determine whether it is physically possible to obtain miscibility with the dry gas of composition given by Point B. If not, how much would it have to be enriched with intermediates (LPG) to do this, in Bbl LPG per MSCF dry gas? The diagram is on a liquid volume basis, and 1000 SCF of methane occupies 0.4 Bbl when dissolved. (Actually the gas available for Block 31 had sufficient inter-mediates.)

2. If the average pressure in the gas zone were 3500 psia, Z for the gas 0.9, from the compressor plant capacity what production rate, STB/D, could be maintained without loss of pressure, assuming only solution gas pro-duced. What would be the life of the field at this rate, assuming that 60% of the oil in place could be recovered?

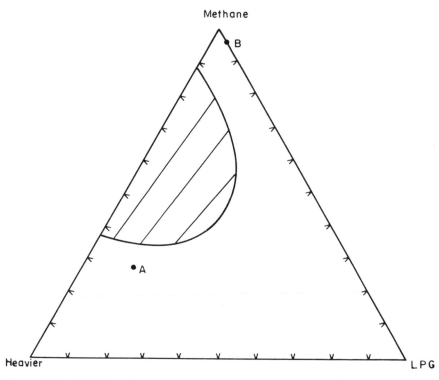

Figure A.1. Enriched (condensing) gas drive, Seeligson Zone.

3.   Estimate the horsepower of the compressor plant (use isothermal formula, Z constant) if the gas was available at 500 psia.

4.   Assume p_w = 4000 psia, p_e = 3500 psia, r_w = 0.333 feet, r_e = 1000 ft. effective well penetration 100 feet, gas viscosity 0.018 cp, permeability and temperature as given in Chapter 1. What is the theoretical intake capacity of such a well?

5.   Sketch the well arrangement for an inverted nine-spot (one injection to three producing wells). What is the theoretical horizontal sweep efficiency at gas breakthrough with the oil and viscosities given (μ_o = 0.3 and μ_g = 0.018)? (Assume a relative permeability ratio of unity).

     Show in the sketch the gas-oil interface at breakthrough. Discuss other factors which would determine the ultimate economic per cent recovery. From the expected ultimate recovery (175 MMSTB) does it seem that a complete miscible drive was obtained?

Answers

1.   0.2 Bbl LPG per MSCF of dry gas.

2.   26 M STB/D, 17 years.

3.   5500 HP.

4.   3 MM SCF/D.

Problem X: Enriched (Condensing) Gas Drive, Seeligson Zone

1.   Assuming no gas cap, find the amount of oil originally in place, using the volumetric data given and taking 1.43 as the original formation volume factor. Is a small cap indicated?

2.   At surface temperatures, (such as in the transmission lines) and several hundred pounds pressure, a 50:50 molal mixture of dry gas and propane would form a liquid and gaseous phase. Yet at 3000 psi, the pressure needed for injectionn, and the same temperature range, there is only one phase. Explain.

3.   Note that 0.2 hydrocarbon pore volume of enriched (50 mole % propane) gas had been injected from March, 1957, to September, 1960. Assuming

2700 psia and 171° F, and a supercompressibility factor Z of 0.5, how many SCF were injected? How many per day? If liquid propane has 0.6 the density of water, how many barrels of propane were injected? What fraction of the hydrocarbon pore volume is this? How many reservoir barrels of gas were injected per day? How many reservoir barrels of oil were produced per day (1.37) during the period 1958-63 if production was 1000 STB/D? (Note that the pressure rose).

4. By Stiles' method, assuming a linear system and, as suggested, 20 feet each of 200 and 75 md sand, what would be the vertical breakthrough sweep efficiency? (To compensate for mobility ratios, assume the 200 md layer to be 400 md). If the volumetric breakthrough efficiency was 21% and the vertical as just calculated, what was the horizontal sweep efficiency?
The statement "75 per cent of the reservoir volume was behind the break-through front" apparently means that 75% of the more permeable layer was swept. At 60% vertical sweep efficiency, what was the volumetric sweep efficiency?

5. Considering that a million barrels of propane and a million barrels of ethane had to be bought, gas compressed, etc., discuss the economic factors that determine whether this operation was more profitable than a water drive, aside from the fact that the estimated ultimate recovery by water drive was somewhat less.

Answers

1. 7.5 MMSTB. Yes.

3. 3.7 MMMSCF; 3 MMSCF/D; 1 MMBbl; Ca. 10%; 1750 res. Bbl. gas per day; 1370 Res. Bbl. oil per day.

4. 60%; 35%; 45%.

Appendix B

Extension of the Welge Tangent Method
To Account for a Uniform and Mobile
Initial Displacing Phase Saturation*

Total volume of gas injected $= E + B + D = \phi A \int_{S_{g_c}}^{S_{g_{max}}} x \, dS_g$ and from

the frontal advance **Equation 11**, section 3.1. **B-1**

$$x = \frac{Q}{\phi A} \frac{df_g}{dS_g} \qquad \text{B-2}$$

where $\quad Q = \int_o^t q \, dt.$

Substituting **Equation B-2** in **Equation B-1**, gives

$E + B + D = Q$ **B-3**

*Kern, L. R.: "Displacement Mechanism in Multi-Well System." *Trans. AIME* (1952) 195, 91.

Now, the volume of gas D flowing out of the swept area is

$$D = \phi A \int_{S_{gc}}^{S_{gi}} x \, dS_g \quad = Q f_{gi} \tag{B-4}$$

and from Figure B.1

$$E + C = (S_{gf} - S_{gi}) \, x \, \phi A + \phi A \int_{S_{gf}}^{S_{gmax}} x \, dS_g$$

$$= (S_{gf} - S_{gi}) \, Q f_{gf}' + Q \, (1 - f_{gf}) \tag{B-5}$$

where the prime denotes differentiation with respect to saturation and the subscript f refers to the front. But

$$E + C = Q - D = Q \, (1 - f_{gi}) \tag{B-6}$$

whereupon equating **Equations B-5** and **B-6**, gives

$$S_{gf} = S_{gi} + \frac{f_{gf} - f_{gi}}{f_{gf}'} \tag{B-7}$$

which shows that the Welge tangent must be drawn from the initial conditions f_{gi}, S_{gi}.

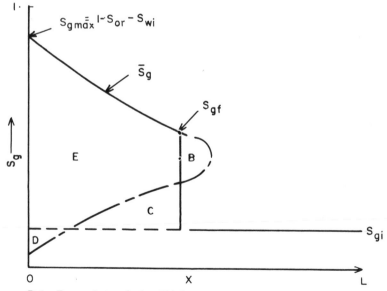

Figure B.1. Extension of the Welge tangent method to account for a uniform and mobile initial displacing phase saturation.

Index

Author Index

Subject Index